# ACIDIFICATION OF FRESHWATERS

CAMBRIDGE ENVIRONMENTAL CHEMISTRY SERIES

*Editors:*

R. M. Harrison, *Department of Chemistry, University of Essex*

G. R. Helz, *Department of Chemistry, University of Maryland*

P. S. Liss, *School of Environmental Sciences, University of East Anglia*

*The first textbook in this series is:*

P. Brimblecombe *Air Composition & Chemistry*

# ACIDIFICATION OF

○ ○ ○ ○ ○ ○ ○ ○ ○ ○ ○ ○ ○ ○ ○ ○ ○ ○

# FRESHWATERS

## MALCOLM CRESSER

*Department of Soil Science, University of Aberdeen*

## ANTHONY EDWARDS

*Department of Soil Fertility, Macaulay Institute for Soil Research, Aberdeen*

The right of the
University of Cambridge
to print and sell
all manner of books
was granted by
Henry VIII in 1534.
The University has printed
and published continuously
since 1584.

## CAMBRIDGE UNIVERSITY PRESS

*Cambridge*

*London   New York   New Rochelle*

*Melbourne   Sydney*

CAMBRIDGE UNIVERSITY PRESS
Cambridge, New York, Melbourne, Madrid, Cape Town, Singapore,
São Paulo, Delhi, Dubai, Tokyo, Mexico City

Cambridge University Press
The Edinburgh Building, Cambridge CB2 8RU, UK

Published in the United States of America by Cambridge University Press, New York

www.cambridge.org
Information on this title: www.cambridge.org/9780521158367

First published 1987
First paperback edition 2010

*A catalogue record for this publication is available from the British Library*

*Library of Congress Cataloguing in Publication Data*

Cresser, Malcolm S.
Acidification of freshwaters.

(Cambridge environmental chemistry series)
Includes bibliographies and index.
1. Acid pollution of rivers, lakes, etc.
I. Edwards, Anthony. II. Title. III. Series.
TD425.C661987   628.1'68   86-26907

ISBN 978-0-521-32270-6 Hardback
ISBN 978-0-521-15836-7 Paperback

# CONTENTS

○ ○ ○ ○ ○ ○ ○ ○ ○ ○ ○ ○ ○ ○ ○ ○ ○ ○ ○ ○ ○

# PREFACE

The causes of acidification of freshwaters have become a major issue to environmental scientists, industrialists and governments over recent years. This book attempts to provide an insight into the interactions between precipitation, acidifying pollution, plants, soils and waters which regulate water acidity. The complexity of the topic, and much of its fascination, arise from the fact that it can be studied best only by an interdisciplinary approach. Inputs are required from all the main branches of chemistry, but also from soil physics and soil chemistry, hydrology, meteorology, geography, geology, plant physiology, soil microbiology, zoology, and even occasionally from engineering. We have attempted, therefore, to present an integrated approach, drawing upon all these disciplines where appropriate, but with a final account that should be understandable and useful in all of them, and to the general public.

Many people have contributed to the existence of this book in its present form, often unwittingly so. We are grateful to: members of the research group in the Soil Science Department at Aberdeen University, especially Murray Reid, John Creasey, Donald MacLeod, Ute Skiba, Tim Peirson-Smith, Simon Ingram, Bob Rees and Fiona Parker-Jervis; our first-rate technical back-up crew, especially George Wilson, Jim Wallace, Bruce Gordon, Mick Fuller, Fiona Mitchell, Alex Rennie, Alec Inglis and Norman Gray; Elspeth Crombie, for secretarial help second to none; the land owners and keepers who have patiently allowed us to work on their estates; Jan Økland and Rick Battarbee for willing permission to reproduce diagrams; the several scientists who have contributed through discussion to the formulation of our ideas, especially E. A. FitzPatrick (Fitz), Fred Last, Bryan Bache, Mike Hornung, Neil Cape and Colin Neal; Ken Pugh and his colleagues at the

North-East River Purification Board for help with analyses and making available much valuable data; Mr Rendall and Mr Smith of Grampian Water Services, for providing data; Sue and Louise, for patiently stepping over piles of reprints.

Last, but by no means least, we are indebted to MAFF, NERC and the UK Department of the Environment for the financial support that made the work possible at all. We are especially grateful to Bob Wilson of DOE for his valued advice and comments over the years.

*Aberdeen, August 1986*

Malcolm Cresser and
Anthony Edwards

# 1

○ ○ ○ ○ ○ ○ ○ ○ ○ ○ ○ ○ ○ ○ ○ ○ ○ ○ ○ ○

# The importance of freshwater acidification

'British lakes poisoned by acid rain' (*The Observer*, 18 September 1983); 'Acid rain kills fish in Welsh rivers' (*The Observer*, 25 September 1983); and 'Britain's two million tonne chemical warfare onslaught' (*The Guardian*, 22 September 1983). These are typical British newspaper headlines dealing with freshwater acidification during the early 1980s. Proclamations of similar ilk have appeared in much of the European and North American press. In the minds of the media-informed public, acid rain and freshwater acidification have now become almost inextricably linked. To the scientific community, however, the issue is a much more complex one.

Scientific issues generally only attract the attention of the mass media and the general public through emotive appeal rather than through any intrinsic scientific interest. Thus the premature partial or complete destruction (either actual or potential) of human, animal and plant life all serve to catalyse newsworthiness. A further component that adds to the attractiveness of a news item is a high cost factor; someone should be paying (or at least threatened with having to pay) dearly as a result of the events reported. Based upon these criteria, 'acid rain' has all the essential ingredients for successful journalistic exploitation. The victims are fish, freshwater invertebrates and the animals, especially birds, that feed upon them, and trees, other plants, and even some of our architectural heritage. It is often suggested that the culprits are the governments and industrialists who advocate the continued exploitation of fossil fuels with supposedly inadequate clean-up of emissions. The high costs are typified by Sir Walter Marshall's estimate of £4000 million to free the UK Central Electricity Generating Board's emission from sulphur dioxide (Anon., 1983). In more immediately meaningful terms, this corresponds to a 10–15% increase in the cost to the CEGB of generating electricity. This conclusion about projected costs is in line with the detailed appraisal of Highton and Webb (1984), which suggests increases of 1 and 5% in electricity prices in England and Wales for 20 and 50% reductions in sulphur dioxide emissions. Costs in the USA depend upon the clean-up procedure advocated, but estimated capital outlays typically vary between $3 billion for use of the best practical coal washing at major sources burning 'washable' coal to $7–14 billion for retrofitting of flue-gas desulphurisation equipment on selected, very large pollutant sources (Wetstone *et al.*, 1981). Substantial reductions in nitrogen oxide emissions from vehicle exhausts which lead to nitric acid production (see e.g. Likens *et al.*, 1979) would also significantly increase the cost-of-living index.

To environmental scientists such a publicity onslaught may be seen as potentially beneficial in so far as it may encourage investment into environmental research. However, this is only true when the question being asked is: What are the fundamental mechanisms which regulate the acidity of freshwaters, and to what extent may the individual contributions from natural and anthropogenic origins be quantified? This is the question which we are attempting to answer in the present text, at least in so far as the confines of present knowledge allow.

There is however a potential danger in the establishment of research programmes as a result of media pressure, namely that research becomes blinkered, aimed at establishing the validity of a preconceived hypothesis, rather than exploring all conceivable hypotheses in an open-minded fashion. At the end of the day the resulting conclusions from the former approach may well not stand up to close scrutiny because of the questions which remain unanswered.

In the context of freshwater acidification there is a real need to study every aspect of the processes involved in regulation of the short-term (hours to days) and long-term (decades to centuries) time scales for acidification of streams, lakes and other natural waters. It is imperative that natural evolutionary processes are considered alongside the effects of acidifying pollutants of anthropogenic origins and changing patterns of land use. The mechanisms are complex. To understand them fully requires a knowledge of geology and geochemical weathering, soil physical and organic chemistry, pedology, soil physics and hydrology, soil biology, aquatic chemistry, aspects of plant science and forestry, and atmospheric physics and chemistry. Aside from the need for a multidisciplinary approach, the situation is further complicated by the lack of suitable natural controls, especially in field investigations on a

catchment scale. Thus the use of empirical approaches involving artifici-
ally modified environments is often essential. While we do not pretend
to be expert in all of the disciplines listed above, we have made a determined effort to consider each of the interacting components of complete
ecosystems, both in our research and in the preparation of this monograph. Although we do not suggest we have all the answers, we hope to
have at least asked the most appropriate questions. Because of the complexity of the problem under investigation, we feel it is worthwhile to
include in the text a brief but critical discussion of the major techniques
which may be used to probe various components of the water/soil/plant/
atmosphere system.

## Developing awareness of the problems of freshwater acidity

As recently pointed out by Franks (1983), the adverse effects of
'acid rain' have long been recognised. He cited reports by Robert Smith
commenting upon acid rain in the area around Manchester, England, as
long ago as 1852. Severe damage to plant life in the immediate vicinity
of sources of gross acidic pollution, such as some coal-fired power
stations and smelters, is very noticeable. However, the visible
symptoms of freshwater acidification, from whatever cause, are less
immediately obvious to the casual observer. Even when the problem is
chronic, clear lakes and encroaching mosses may not suggest any serious
problem.

With a gradual decline in freshwater pH, some fish may start to
exhibit slight skeletal deformities (Harvey, 1985). Such deformities are
particularly common in the tails of salmon (Campbell, 1985). This will
be followed, with further acidification, by changes in fish age distribution, sometimes with years missing completely, leading to reduced
fish populations and, in extreme cases, total loss of a population, particularly of salmon and trout. On a shorter time scale, fish kill associated
with acidification may sometimes occur to an alarming degree even after
a relatively short single episode of acid water production if the water
chemistry changes dramatically. One such instance which has been well
documented is the loss of *Salmo trutta* from the Tordal River in southern
Norway in the spring of 1975 (Leivestad and Muniz, 1976). Thousands
of dead trout littered the bottom of a 30-km stretch of the river. The pH
fell to 4.65 in the river at early snowmelt, and values as low as 4.0 were
recorded in some tributaries.

It should be stressed here that it is an over-simplification to think that only acidity needs to be considered in the context of fish kills. Low pH waters in rivers often may contain elevated levels of aluminium, which may itself pose a threat to biological communities (see e.g. Driscoll *et al.*, 1980), and low levels of calcium, which may further exacerbate the problem (e.g. Brown, 1982). The significance of these points is discussed briefly later in this chapter, and the reasons for their occurrence in Chapter 2. Declining fish populations tend to be more noticeable and to have more emotive impact than reductions in numbers (or elimination) of some of the smaller invertebrates and macro and microflora. However, the latter groups are of comparable ecological significance. Reduction in food sources may obviously affect higher-mammal populations simply by breaking a link in the food chain. Birds may be influenced in this way for example. If the acid rain cloud has an ecological research silver lining, it must be that the relics of changing diatom populations with depth in lake sediment cores may be used to provide very valuable information as to the rate of change of lake-water pH over past centuries (e.g. Battarbee *et al.*, 1985).

Old records of fish catches are an important source of circumstantial evidence for the effects of water acidification since the turn of the century. Where records have been kept over the past two to three decades for waters thought to have become more acidic by 1–2 pH units or more, depletion of fish stocks would indeed appear to suggest cause for concern. For example, the pH of Lake Lumsden, in the La Cloche Mountains, Southern Ontario, Canada, reputedly fell from 6.8 in 1961 to 4.4 in 1971, with the gradual total loss of all eight fish species over the same time scale (Harvey, 1980). A study of 40 lakes in the Adirondacks, New York, apparently demonstrated a pronounced pH decline between 1930 and 1975, the median pH value falling by 1.6 units to 5.1 (Schofield, 1976). About 200 lakes in the Adirondack Mountains are currently fishless (Harvey, 1980). Apparent acidification effects in Europe have been summarised in a report by Environmental Resources Ltd (1983). Of 5000 lakes in southern Norway, 1750 were 'thought' to have lost fish populations and 900 others to be seriously adversely affected. In central and southern Sweden, 2500 lakes exhibited damage to fish stocks. In the UK, several Scottish lochs and many upland British mainland streams were found to be fishless. In many of the above cases there is said to be strong evidence to suggest that substantial acidification has occurred over the past three to four decades (Environmental Resources Ltd, 1983).

Much of the evidence has come from painstaking piecing together of fragmentary evidence from diverse sources, following the stimulating presentations and discussion at the Telemark international conference on the effects of acid precipitation in the mid 1970s. An indication of the exponential growth of interest in acidifying pollutants following this meeting, and the first international symposium on acid precipitation and the forest ecosystem in Ohio at about the same time, is given in Fig. 1.1, which shows the publication date distribution for papers included in a report on acid rain published in 1983.

The accumulating circumstantial evidence suggested that freshwater acidification was a problem which had rapidly worsened over a few decades, both in terms of the area affected and the extent of pH drops observed. The precise reasons for the suddenness of the deterioration were not always immediately obvious, but it was widely assumed that acidifying pollutants in the atmosphere were largely to blame, even although the evidence was generally circumstantial at best. Until the early 1980s, very little attention was paid either to the time scale of natural freshwater acidification processes, which are discussed in Chapter 2, or to reasons why interactions between acidic deposition and soils were no longer apparently capable of coping with the sustained heavy acid loadings being imposed. Over the past five years, significant progress has been made with this aspect. The conclusions which may so far be drawn are discussed in Chapter 3.

Fig. 1.1. Publication date distribution for references in a review of the acid rain phenomenon in Europe (Environmental Resources Ltd, 1983)

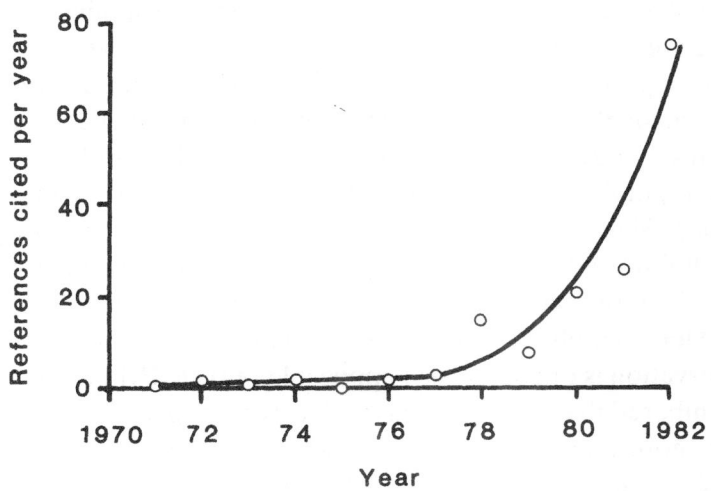

It should be stressed here that a substantial part of the 'evidence' apparently accumulating in the literature must be interpreted with great caution, a generally valid criticism for historical reconstructions (Kramer and Tessier, 1982). The pH of stream water, and of some lake waters, may vary with time of year of sampling, climate prior to sampling, climatic change from year to year, changes in land use and soil management practices, and so on (see Chapter 2). Such factors may have been totally ignored, or at best inadequately considered, in some studies, and data based upon spot samples at an interval of a decade or more should be treated with caution. A long run of pH and alkalinity data is more desirable, but even then possible errors from changes in analytical methodology must be borne in mind. Even today the accuracy of measurement of pH in relatively pure freshwaters is a cause for concern. In countries for which fishless lakes are cited by the thousand, it is important to consider carefully how sure we can be that those lakes supported healthy fish populations a few decades ago. All too often natural acidification processes are simply ignored. One point about which there can be little doubt is that the acid-deposition related research of the past decade has led to a growing awareness of the extent of freshwater acidification, whatever the cause. To some extent the upward trend in reported numbers of fishless lakes should be seen as a reflection of this increased awareness.

## Acidification and water quality

Aside from effects of freshwater acidity upon aquatic biota which have already been mentioned, it is also appropriate to consider briefly here possible effects of acidification upon water supplies for human consumption. A report on the susceptibility of UK groundwaters to acid-deposition-induced acidification (Kinniburgh and Edmunds, 1984) made the point that the most serious consequence of a significant fall in groundwater pH was likely to be the increased solubility of metals in potable water and accelerated corrosion of the distribution network. Since, however, UK groundwaters were generally alkaline and well-buffered, and most supplies of any size were in any case treated to minimise such risks, the greatest threat was probably to small, untreated local supplies from springs and shallow wells and boreholes. This observation is of course also applicable outside Britain, but it must be remembered that in some areas such private supplies may be relatively much more important. Acid groundwaters in shallow wells and at lake inflows have been reported in Sweden (Hultberg and

Johansson, 1981). Deep groundwaters are apparently not acidified in either Norway or Sweden (Henrikson and Kirkhusmo, 1982). Surface waters also are generally treated before being used as major supplies, and here, too, acidification itself is not generally regarded as a serious quality control problem.

It is appropriate also to consider briefly some of the other water quality parameters associated with acidic waters, even if these may be changing coincidentally with acidity rather than as a result of hydrogen ion concentration changes. The dissolved organic matter content of water in upland streams in moorland or afforested areas with a substantial excess of precipitation over evapotranspiration (see Chapter 2) tends to increase substantially as discharge rises during heavy storm events. Concurrently, as an increasing portion of water drains through organic surface horizons, its pH tends to fall temporarily. Both effects are due to hydrological changes; water draining laterally through the more acidic, surface, organic horizons making a greater relative contribution to the discharge. Thus flushes of acidity are often associated with high total organic carbon content (TOC) (see e.g. Edwards *et al.*, 1985a). On the other hand, as precipitation acidity increases, the TOC in water draining vertically from organic surface horizons tends to decrease, possibly because of suppression of organic acid dissociation (McColl *et al.*, 1982, Skiba and Cresser, 1986). Thus the long-term trend in water colour as a result of freshwater acidification may be a decrease, unless the depth of organic surface horizons increases. Acid flushes will still be associated with high TOC unless the acid flush is related to *direct* snow inputs. Meltwater flushes may still be associated with an elevated TOC. These aspects are discussed more fully in Chapters 2 and 3. Certainly many acid lakes have very clear water.

There is much evidence to suggest that acid flushes may be associated with high stream-water concentrations of sulphate and nitrate. This is particularly the case during snowmelt, when the processes which normally help to regulate the drainage of these species into streams and lakes may be substantially curtailed (see e.g. Edwards *et al.*, 1985b). For nitrate the problem may be further exacerbated by the soil freeze/thaw effect, which may mobilise even more nitrate (Likens *et al.*, 1977; Edwards *et al.*, 1986). Ultimately, high nitrate levels could present more of a water quality problem than high hydrogen ion concentrations. There is strong historical evidence for a dramatic increase in the nitrate component of acid rain in Europe and North America (Likens *et al.*, 1977; Brimblecombe and Stedman, 1982). When high concentrations of sulphate pass through soils to drainage water, there is often an associated increase in concentrations of base element and other metal cationic species. Aluminium mobilised in this way may play a major role in the adverse effects of acidic freshwaters on fish and other aquatic life. Sulphate anion adsorption leads to base cation retention, as discussed later, but sulphate saturation of adsorption sites facilitates cation leaching (Cresser *et al.*, 1986).

## Mechanisms of acidification effects on freshwater biota

Before examining the factors contributing to or regulating the extent of freshwater acidification, it is appropriate to consider the mechanisms by which biota are adversely affected. It is important to know what, if any, other factors must be considered simultaneously when assessing the potential impact of acidification. A useful summary of the effects of acidification on fish has been published by Harvey (1985). Apart from the population changes already described, changes in age-class composition and growth and condition are often observed. In the former instance a shift to older specimens is most common, as a result of maturation failure, spawning failure or death of larval fishes, or a combination of two or more of these effects. In some instances, however, the reverse trend has been reported; the population becomes excessively juvenile as a result of death at maturity. Both increases and decreases in size at a given age may be observed, increases presumably resulting from reduced competition with population declines. Skeletal deformities are also reported from time to time, as mentioned earlier.

Retention of fish in acidic water causes loss of ions from plasma and muscle, 35–40% loss of plasma $Na^+$ and $Cl^-$ coinciding with death of rainbow trout from Plastic Lake in Ontario (Harvey, 1985). At the same time aluminium on the gills increased from 20 to 100 $\mu g\,g^{-1}$ gill dry mass. The same study showed elevated manganese (two- to five-fold) in spinal bone. Muniz and Leivestad (1980a) commented that it seemed proved beyond doubt that failure in body salt regulation was the primary cause for fish death in acid water. They concluded that, since calcium and sodium tended to relieve the stress induced by hydrogen ions, fish should be able to tolerate a lower pH in lakes with higher concentrations of these ions. Brown (1982) also found that elevated calcium ameliorated to some extent the adverse effects of low pH. Grouping data for Norwegian lakes on the basis of salinity verified that barren lakes were more frequent where the water had very low ionic contents. Muniz and

Leivestad also concluded that acidified lake and river water contained species other than hydrogen ions that were toxic, because tap water acidified to a similar pH was less toxic. Subsequent investigations demonstrated that inorganic aluminium also caused rapid loss of plasma salts even when present at concentrations below 1 µg ml$^{-1}$. Massive mucus clogging of the gills, hyperventilation, coughing and respiratory stress were typically observed in addition to ion loss (Muniz and Leivestad, 1980b). Exposure to dissolved inorganic aluminium species may be toxic to fish at pH values of 5.5 or higher, even for relatively short periods (Grahn, 1980; Muniz and Leivestad, 1980a), another contributing factor to the observation that particular species populations may be eliminated at different pH values in different lakes. Aluminium which is not organically complexed tends to be more toxic than that which is (Driscoll *et al.*, 1980). Thus it appears that, when considering freshwater acidification in relation to fish kills, it is necessary to consider not just pH, but also salinity and inorganic monomeric aluminium concentration.

Apart from the direct toxicity effects upon fish discussed above, it is also necessary to remember that hatching of eggs may be delayed, or even totally suppressed, at low pH for sensitive species such as the Atlantic salmon (Peterson *et al.*, 1980). A further consideration is the adverse impact which acidification may have upon the availability of food. For example, Økland and Økland (1980) found that the freshwater shrimp, *Gammarus lacustris*, a major food for trout in Norwegian lakes, was invariably absent from upland lakes below pH 6.

The results of a survey of 1500 freshwater sites in Norway summarised by Økland and Økland (1980) also showed that snails, another important fish food source, were not generally found below pH 6, although isolated species occurred down to pH 5.2 at low population densities. The behaviour of small mussels was rather similar, except that some survived down to pH 4.7. The freshwater louse was generally more acid tolerant, being widely found down to pH 5.2, but less frequently in waters down to pH 4.8. Fig. 1.2 shows the lower pH tolerance limit for 17 species of snails, mussels and crustaceans which are widely found in Norwegian waters close to neutrality. Raddum *et al.* (1980) have studied the numbers of zooplankton species, including Crustacea, Rotifera and *Chaoborus* (a planktonic Diptera), in acid and non-acid lakes in southern and western Norway. Species diversity was much reduced in the acidified lakes, but less so when the acid was organically derived to a large extent. The same group also studied phytoplankton. They found no correlation between chlorophyll *a* content and pH, although the phaopigment contents showed substantial variability. The number of algal species in the phytoplankton was, at a given time, generally lower in the acid lakes. It was concluded that high acidity in combination with high dissolved organic matter affected species distribution both through the tolerance of individual species and also by changing the dynamics of species succession and inter-species interactions.

The importance of the work discussed in the preceding few pages and the substantial body of similar work is invaluable in establishing the nature of the link between acidification of freshwater and the biological species supported. It should perhaps be stressed, however, that much of

Fig. 1.2. Lower pH tolerance limit for 17 common species of snails, mussels and crustaceans found in Norwegian near-neutral waters. After Økland and Økland (1980), with permission of the authors.

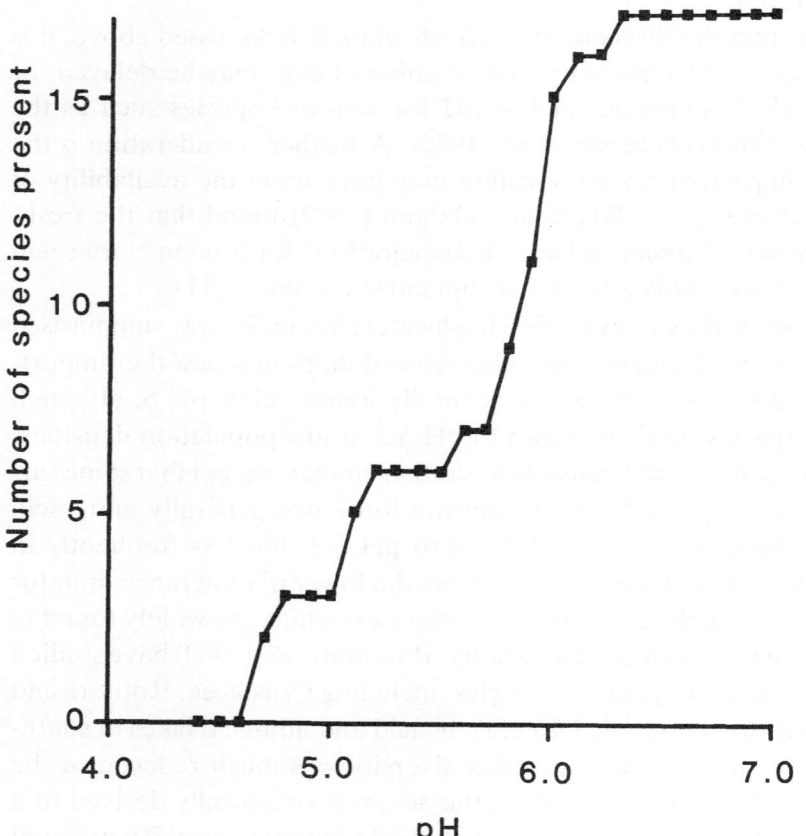

the research published does not attempt to impiicate atmospheric pollution, but rather studies acidification effects from whatever cause. As will be seen in later chapters, heavy sulphate inputs may lead to calcium leaching from soil at depth, thus increasing drainage water calcium concentrations. As Brown (1982) has pointed out, this may ameliorate the effect of the acidity to some extent, at least in the medium term. On the other hand, drainage water from naturally acidic upland soils may be most acidic when calcium concentration is at its lowest, potentially a more serious problem. It is imperative, therefore, to study and understand the natural, unavoidable acidification processes before any irrefutable observation may be made about the effects of atmospheric acidifying pollutants. Thus naturally occurring processes are discussed first in this text. At times there is inevitably overlap between the discussion of these processes and that of the effect of pollutants of anthropogenic origins. Repetition has been kept to a minimum by cross-reference where appropriate. However, the reader requiring a balanced overview of the problem is strongly encouraged to read Chapter 2 before Chapter 3.

We have not attempted to provide a comprehensive review of the impact of freshwater acidification on biota here, but rather to give some insight into the very real cause for concern. Those requiring more information should consult some of the references cited, or the useful select bibliography *Acid Rain and the Environment* and the associated updates compiled and edited by Farmer (1984).

## References

Anon. (1983) Scandinavian acid rain: Royal Society appointed referee in UK dispute. *Nature* (London), **305**, 85.

Battarbee, R. W., Flower, R. J., Stevenson, A. C. and Rippey, B. (1985) Lake acidification in Galloway: a palaeoecological test of competing hypotheses. *Nature* (London), **314**, 350–2.

Brimblecombe, P. and Stedman, D. H. (1982) Historical evidence for a dramatic increase in the nitrate component of acid rain. *Nature* (London), **298**, 460–1.

Brown, D. J. A. (1982) The effect of pH and calcium on fish and fisheries. *Water, Air and Soil Pollution*, **18**, 343–51.

Campbell, R. N. B. (1985) An acidification study of Scottish lochs, paper presented at the British Ecological Society Meeting, Edinburgh, December 1985. *British Ecological Society Bulletin*, **16** (4), 229.

Cresser, M. S., Edwards, A. C., Ingram, S., Skiba, U. and Peirson-Smith, T. (1986) Soil-acid deposition interactions and their possible effects on geochemical weathering rates in British uplands. *Journal of the Geological Society, London*, **143**, 649–58.

Driscoll, C. T., Baker, J. P., Bisogni, J. J. and Schofield, C. L. (1980) Effects of aluminium speciation on fish in dilute acidified waters. *Nature* (London), **284**, 161–4.

Edwards, A. C., Creasey, J., Skiba, U., Peirson-Smith, T. and Cresser, M. S. (1985a) Long-term rates of acidification of UK upland acidic soils. *Soil Use and Management*, **1**, 61–5.

Edwards, A. C., Creasey, J. and Cresser, M. S. (1985b) Factors influencing nitrogen inputs and outputs in two Scottish upland catchments. *Soil Use and Management*, **1**, 83–7.

Edwards, A. C., Creasey, J. and Cresser, M. S. (1986) Soil freezing effects on upland stream solute chemistry. *Water Research*, **20**, 831–4.

Environmental Resources Ltd (1983) *Acid Rain: A Review of the Phenomenon in the EEC and Europe.* Graham and Trotman Ltd, London, pp. vii + 159.

Farmer, P. (1984) *Acid Rain and the Environment 1980–1984 – A Select Bibliography.* Technical Communications, Letchworth, England, pp. 108.

Franks, J. (1983) Acid rain. *Chemistry in Britain*, June, 504–9.

Grahn, O. (1980) Fish kills in two moderately acid lakes due to high aluminium concentration. In *Ecological Impact of Acid Precipitation*. Drabløs, D. and Tollan, A., eds, SNSF Project, Oslo, 310–11.

Harvey, H. H. (1980) Widespread and diverse changes in the biota of North American lakes and rivers coincident with acidification. In *Ecological Impact of Acid Precipitation*, Drabløs, D. and Tollan, A., eds, SNSF Project, Oslo, 93–8.

Harvey, H. H. (1985) The biological evidence of acidification, mechanism of action and an attempt at predicting acidification effects. In *Symposium on the Effects of Air Pollution on Forest and Water Ecosystems, Helsinki, April 23–24, 1985*, Foundation for Research of Natural Resources in Finland, Helsinki, 63–78.

Henrikson, A. and Kirkhusmo, L. A. (1982) Acidification of groundwater in Norway. *Nordic Hydrology*, **13**, 183–92.

Highton, N. H. and Webb, M. G. (1984) The effects on electricity prices in England and Wales of national sulphur dioxide emission standards for power stations. *Journal of Environmental Economics and Management*, **11**, 70–83.

Hultberg, H. and Johansson, S. (1981) Acid groundwater. *Nordic Hydrology*, **12**, 51–64.

Kinniburgh, D. G. and Edmunds, W. M. (1984) *The Susceptibility of UK Groundwaters to Acid Deposition – Report to the Department of the Environment*, British Geological Survey, Wallingford, pp. 211.

Kramer, J. and Tessier, A. (1982) Acidification of aquatic systems: A critique of chemical approaches. *Environmental Science and Technology*, **16**, 606A–615A.

Leivestad, H. and Muniz, I. P. (1976) Fish kill at low pH in a Norwegian river. *Nature* (London), **259**, 391–2.

Likens, G. E., Bormann, F. H., Pierce, R. S., Eaton, J. S. and Johnson, N. M. (1977) *Biogeochemistry of a Forested Ecosystem.* Springer-Verlag, New York.

Likens, G. E., Wright, R. F., Galloway, J. N. and Butler, T. J. (1979) Acid rain. *Scientific American*, 214 (4), 43–51.

McColl, J. G., Browe, A. B. and Firestone, M. K. (1982) Acid rain mobilization of aluminium and organic carbon. *Agronomy Abstracts*, 1982, 178.

Muniz, I. P. and Leivestad, H. (1980a) Acidification-effects on freshwater fish. In *Ecological Impact of Acid Precipitation*, Drabløs, D. and Tollan, A., eds, SNSF Project, Oslo, 84–92.

Muniz, I. P. and Leivestad, H. (1980b) Toxic effects of aluminium on the brown trout,

*Salmo trutta* L. In *Ecological Impact of Acid Precipitation*, Drabløs, D. and Tollan, A., eds, SNSF Project, Oslo, 320–1.

Økland, J. and Økland, K. A. (1980) pH level and food organisms for fish: Studies in 1000 lakes in Norway. In *Ecological Impact of Acid Precipitation*, Drabløs, D. and Tollan, A., eds, SNSF Project, Oslo, 326–7.

Peterson, R. H., Daye, P. G. and Metcalfe, J. L. (1980) The effects of low pH on hatching of Atlantic salmon eggs. In *Ecological Impact of Acid Precipitation*, Drabløs, D. and Tollan, A., eds, SNSF Project, Oslo, 328.

Raddum, G. G., Hobaek, A., Løuisland, E. R. and Johnsen, T. (1980) Phytoplankton and zooplankton in acidified lakes in South Norway. In *Ecological Impact of Acid Precipitation*, Drabløs, D. and Tollan, A., eds, SNSF Project, Oslo, 332–3.

Schofield, C. L. (1976) Acid precipitation effects on fish. *Ambio*, **5**, 228–30.

Skiba, U. and Cresser, M. S. (1986) Effects of precipitation acidity on the chemistry and microbiology of Sitka spruce litter leachate. *Environmental Pollution* (A), **42**, 65–78.

Wetstone, G., Reed, P. and Futrell, J. W. (1981) Alternatives for coping with acid deposition problems. *Environmental Policy and Law*, **7**, 155–8.

# 2

○ ○ ○ ○ ○ ○ ○ ○ ○ ○ ○ ○ ○ ○ ○ ○ ○ ○ ○ ○ ○

## Natural acidification processes

It is important to realise from the outset that numerous ecosystems exist in which drainage water may be naturally very acidic. In the present context, 'very acidic' implies a pH value of 4.5 or less. These low-pH waters would still exist even in the complete absence of acidifying pollutants of anthropogenic origins in the atmosphere. The purpose of this chapter is to explain the mechanisms involved in these natural water acidification processes, and to give some insight into the environmental conditions under which such acidified waters are likely to be found. Because much of the water draining into streams or lakes passes through soil to a greater or lesser extent, it is necessary to consider also the processes of soil formation and pathways of water movement which regulate soil acidity, and the interactions between precipitation and the plant/soil system. Only when the natural processes are understood is it reasonable to make an assessment of the possible effects of acidifying pollutants and of human activities in general.

The most important single factor in regulation of drainage water acidity is undoubtedly the circulation of water, more especially that water which flows through and/or over soils to streams or lakes or enters lakes directly. Discussion of the many facets of water movement is therefore the most appropriate point at which to start this chapter.

### The hydrologic cycle

A simplified schematic representation of the hydrologic cycle is shown in Fig. 2.1. By far the largest percentage (94.2%) of the world's total water constitutes the oceans (L'Vovich, 1979). Groundwater and ice sheets contain approximately 4 and 1.5% of the total amount, respectively. Significant redistributions between these three major components occur over a time scale of centuries to tens of centuries. The water which is of most relevance in the present context is only a fraction of one per cent of the total water reserve. Redistribution between the components that make up this fraction, namely precipitation (rain, mist, snow, etc.), atmospheric water vapour and cloud, sea spray in the atmosphere, soil water and river and lake waters, occurs over a much shorter time scale, days and weeks rather than centuries.

The frequency and intensity of individual rain events varies considerably, both with respect to geographical location and also seasonally in many instances. In the northern hemisphere, a substantial quantity of the annual precipitation amount is closely associated with relief and dominant frontal patterns. This is the case for example in the uplands of western Britain, the USA, Canada, Sweden and Norway, i.e. in the areas where freshwater acidification is recognised to be occurring. Obviously certain local variations are superimposed upon the global scheme. There is also a growing awareness of cyclic trends in precipitation amounts in some areas, which can range over periods of two to three decades (see e.g. Vines, 1984). Climatic fluctuations, of both temperature and precipitation characteristics, also tend to occur over the time scale of a few centuries or longer. As will be seen later, these

Fig. 2.1. A simplified schematic representation of the hydrologic cycle.

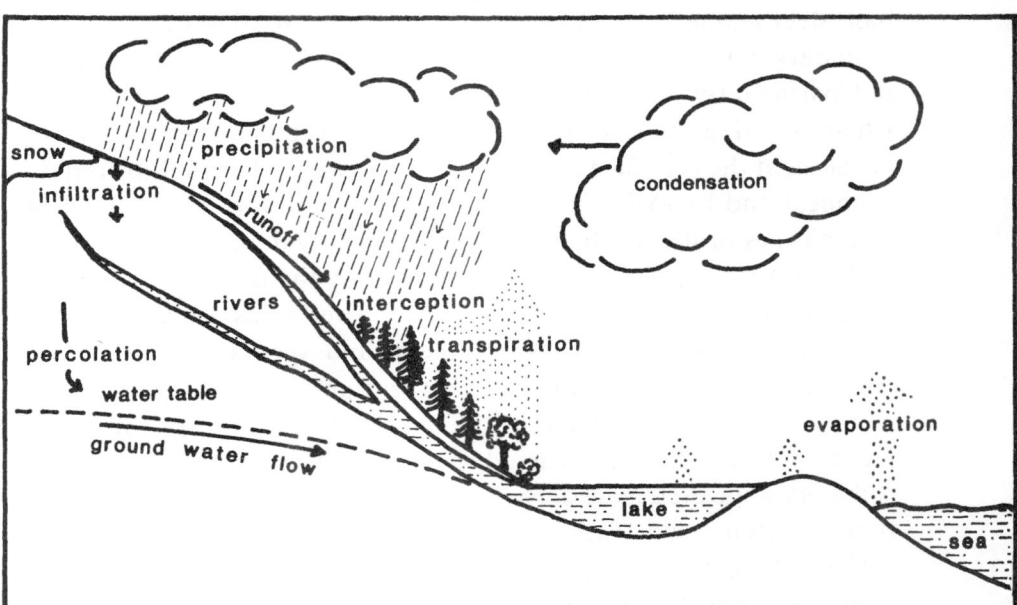

natural cyclic changes in the patterns of precipitation (distribution and quantity) may significantly influence the hydrologic pathways of water flowing to a stream or lake, and thus potentially could modify water chemistry. Where such medium and long-term cyclical trends exist, they must be taken into account while interpreting medium to long-term changes occurring in freshwater pH. It should be remembered that they may also have indirect effects, for example, via vegetational changes, as well as more direct effects.

## Atmospheric inputs to ecosystems

Atmospheric inputs to ecosystems are placed for convenience into one of two broad categories, wet deposition or dry deposition. Dry deposition is the term used to describe the collection and retention (temporary or more permanent) of aerosol or gaseous components of the atmosphere by a variety of physical processes on vegetation, water, and/or soil surfaces. 'Aerosol' is used here to describe any solid particulate matter transported through the atmosphere. The term is sometimes also taken to include suspended solution droplets which are sufficiently small not to be susceptible to gravitational settling at a significant rate. Wet deposition inputs are solute inputs which occur as a result of precipitation of atmospheric liquid or solid water on to rock, soil, water or vegetation surfaces. Wet deposition thus includes inputs in rain, snow, hail, sleet and mist. It will be seen from the above description that the boundary between what precisely constitutes wet and dry deposition is not particularly sharp. However this does not present any problems when assessing the relevance of deposition to freshwater acidification because, at the interface between the two groups, the resulting physico-chemical and biological processes occurring are effectively the same. The physics of the mechanisms involved in dry deposition is complex, and it is not yet possible to quantify reliably the processes for diverse vegetation types under different climatic conditions. Obviously, however, factors such as surface wetness, and vegetation size, growth patterns and surface roughness are important. The topic of dry deposition arises again in Chapter 3, when the significance of forest plantation and clearance is discussed. It should be remembered that, while some dry deposited material may be washed off in a subsequent storm, there is often a substantial period of contact time with the vegetation surface. Thus there may be plenty of time for uptake or chemical or physico-chemical reactions such as ion exchange at the leaf surface if sufficient moisture is available, as for example in misty conditions (Skiba *et al.*, 1986).

Precipitation (rain, hail, snow, sleet) provides a major mechanism for transfer of atmospheric inputs to the terrestrial system. Solute may be acquired by two processes, depending upon where in the atmosphere scavenging occurs. Rainout is the term applied to inputs which originate in the original cloud system, while washout describes the removal of material by falling precipitation. Dry deposition inputs may sometimes be comparable to, or more important than, wet deposition loadings. Geographical location is one of the key factors regulating the relative importance of the two processes, with wet deposition generally making the major contribution in upland regions with higher rainfall amounts (Ottar, 1978; Fowler *et al.*, 1985). Seasonal differences may, however, sometimes occur in this pattern, for reasons discussed later.

Atmospheric inputs of chemical species of interest may originate from one or more of a variety of sources. These include oceanic spray, wind-blown terrestrial dust (of both biological and mineral origins), and gaseous and particulate emissions arising both as a result of human activities (e.g. fossil fuel combustion, intensive animal rearing, quarrying, spreading of lime and fertilisers, etc.) and also natural processes (e.g. volcanic activity, forest fires, gas production under anaerobic conditions, animal excreta, oceanic production of dimethylsulphide, etc.). Interactions within the atmosphere may result in substantial modification of chemical species after they have entered the atmosphere. The brief discussion of inputs presented below has not been confined to hydrogen ions ($H^+$) in isolation, because, as will become clear later, many other facets of precipitation chemistry may be equally or even more important in the context of freshwater acidification. Chemically and biologically controlled reactions within the lithosphere produce their own sources of acidifying and/or neutralising compounds which, in many instances, mean that atmospheric $H^+$ inputs to the soil/plant system are of little consequence to water acidity in the short term. To some of these reactions, inputs of species other than $H^+$ may be much more important.

Often both naturally occurring and anthropogenically derived material present in the atmosphere originates from point sources. During heavy precipitation this may lead to very localised high solute concentrations. At other times such material may disperse over very considerable distances (i.e. on a global scale). The mechanisms of transport and the atmospheric residence times of species are therefore extremely important in regulating atmospheric inputs at any specific point. Geographical location and dominant wind direction are thus particularly significant, and precipitation chemistry may be extremely variable, even over distances of a few kilometres. For this reason, and also because of the dramatic changes which may occur in precipitation solute composition even over the period of a single storm as a result of changes in sub-cloud scavenging (e.g. Edwards *et al.*, 1984), precipitation chemistry often tends to be much more variable than river-water chemistry in a given catchment, the latter generally being far better buffered against change by solute–solid interactions in the soil. Table 2.1, for example, shows typical coefficients of variation for various species in rain and river water.

## Natural materials deposited from the atmosphere

It is appropriate here to consider briefly the major sources of precipitation solute and of the particulates and gaseous components that participate in dry deposition. In many instances a particular source contributes to both wet and dry deposition. Emphasis is placed at this point upon naturally derived (rather than anthropogenic-activity derived) species, although it is sometimes difficult to quantify inputs from the separate sources. It is intended that this section should give some insight into how and why precipitation and dry deposition chemistry might be expected to vary at a given site. The significance of the dis-

Table 2.1. *Average concentrations (mg $1^{-1}$) and coefficients of variation (%) of chemical species in weekly samples of waters in the Grampian Region, 1 October 1977 to 30 September 1978*

| Species | Precipitation | | River Dye | | River Dee | | River Don | |
|---|---|---|---|---|---|---|---|---|
| | Mean | C.V. | Mean | C.V. | Mean | C.V. | Mean | C.V. |
| $H^+$ | 0.058 | 47.2 | | | | | | |
| $Ca^{2+}$ | 0.47 | 70.2 | 3.09 | 30.8 | 5.43 | 40.6 | 11.85 | 10.7 |
| $Mg^{2+}$ | 0.22 | 77.1 | 1.14 | 21.2 | 1.35 | 35.4 | 3.02 | 11.9 |
| $Na^+$ | 1.83 | 86.4 | 5.18 | 13.6 | 4.55 | 28.6 | 7.90 | 10.8 |
| $K^+$ | 0.18 | 108.7 | 0.44 | 23.0 | 0.59 | 50.7 | 1.50 | 28.2 |
| $HCO_3^-$ | 0 | — | 10.67 | 67.1 | 13.17 | 34.4 | 34.64 | 24.5 |
| $Cl^-$ | 2.93 | 84.9 | 6.12 | 8.2 | 6.78 | 42.7 | 12.35 | 14.5 |
| $SO_4^{2-}$–S | 1.56 | 33.8 | 2.25 | 19.2 | 2.61 | 27.0 | 3.85 | 14.6 |

Reid, J. M., unpublished results

cussion should become clear later when soil/water interactions are discussed.

*Oceanic aerosol sources*

Oceans provide a major source of atmospheric particulates as well as being a major sink for deposited material. Duce and Hoffman (1976) reported that between 30 and 75% of the total 'natural' global particulate production ($<20$ μm) comes from this source, and that *ca* 10% of the salts thus mobilised through the atmosphere are deposited on land. Particularly in countries with a maritime climate, oceanic aerosol is the major source of $Na^+$, $Mg^{2+}$ and $Cl^-$ and a substantial part of the $SO_4^{2-}$ budget. Salts entering the atmosphere as a result of bubble bursting mechanisms at breaking waves do not necessarily exhibit the same composition as the parent sea water, because of the ionic redistribution which may conceivably occur during the aerosol formation (see e.g. Bloch *et al.*, 1966). As might be expected, increased storminess of the oceans during winter tends to lead to increased concentrations of sea-salt derived species in precipitation (Reid *et al.*, 1981).

*Volcanic emissions*

Volcanic eruptions introduce both gaseous and particulate materials into the atmosphere. They are obviously rather episodic by nature, but may have long-reaching after effects. A range of sulphur compounds may be emitted, most being subsequently oxidised to sulphur dioxide, the predominant important gaseous product of volcanic activity (Cadle *et al.*, 1971; Gerlach and Nordlie, 1975). Although individual volcanic emission episodes have been traced in polar ice (Delmas and Boutron, 1978; Delmas, 1979), it has been suggested that the total annual release of sulphur to the atmosphere from this source is probably no greater than 5 Tg $a^{-1}$ (Cullis and Hirschler, 1980). Fine volcanic ash may contribute slightly to the deposition of a range of elements at a substantial distance from the original point source.

*Biogenic sources*

Granat *et al.* (1976) and Cullis and Hirschler (1980) have suggested that biological reduction of sulphur compounds constitutes the most important *natural* source of atmospheric sulphur. This occurs as a result of non-specific reduction in marine algae, soils, and decaying vegetation, and more specific bacterial reduction as is found in muds (Hitchcock, 1976). Sulphur dioxide of biogenic oceanic origin is particularly important, but the gas has a residence time in the marine atmosphere of only about 0.7 days (Nguyen *et al.*, 1978). Dimethylsulphide produced is oxidised rapidly to sulphur dioxide (Lovelock *et al.*, 1972). Granat *et al.* (1976) made the point that hydrogen sulphide was also extensively produced by sulphate-reducing bacteria in lower continental shelf sediments, but was unlikely to escape from deep-water sediments into the atmosphere because of the ease of oxidation in aerated water. They concluded that displacement from estuarine waters was probably more significant. Rice *et al.* (1981) assessed the relative importance of different gaseous sulphur species from diverse sources in parts of the USA, and found that local factors, particularly whether the surrounding area was industrial or rural, were important. They further suggested that the greatest range in rates observed in atmospheric sulphur contents in rural areas indicated that the area of land from which hydrogen sulphide emission was occurring was a key parameter in the regulation of atmospheric gaseous sulphur compound concentration. Hydrogen sulphide is not stable, and is soon oxidised in the atmosphere. Estimates of the rate of oxidation vary, but the 24-h survival time suggested by Cox and Sandalls (1974) is not atypical. Biological regulation of atmospheric sulphur concentration leads to a seasonal (temperature-dependent) effect. Other factors may also be important seasonally. For example, Rice *et al.* (1981) have suggested that sulphurous gas emissions from large lakes such as Lake Erie may be particularly high during spring and autumn, when surface layers are disrupted and water with a low oxygen content comes to the surface as a consequence of mixing.

A number of gaseous nitrogen compounds occur in the atmosphere as a consequence of biological activity. Sea water is a natural source of nitrous oxide (Cohen and Gordon, 1979). There is considerable debate as to the importance and even the origins of the gas from this source, some authors suggesting denitrification as the major source while others believe nitrification is more important. Terrestrial nitrous oxide production, from denitrification in soils, is probably very important. It should be stressed, however, that some of the gaseous components of the nitrogen cycle have been less extensively studied on a quantitative global scale than those of the sulphur cycle in the context of acid deposition. As far as is known, the residence time of nitrous oxide in the atmosphere is very long (several decades), and it plays no apparent part in freshwater acidification (Tabatabai, 1985). Nitric oxide (NO) and nitrogen dioxide ($NO_2$) occur in significant amounts in the atmosphere, but are relatively short lived and may, after oxidation, lead to nitric acid production. Both gases are often predominantly present as a result of pollution rather than biological activity, although they may be produced in natural fires.

Ammonia may be produced by volatilisation from animal excreta, or as a result of microbiological mineralisation of organic nitrogen under alkaline conditions. As discussed at some length in Chapter 3, ammonium ions in rain may be a major factor in forest soil acidification under some circumstances. Much volatilised ammonia reacts with acidifying gases in the atmosphere to produce ammonium salts. Ammonium sulphate, ammonium hydrogen sulphate and ammonium nitrate are common components in atmospheric aerosols (Tabatabai, 1985).

*Wind-blown terrestrial dust*
Unlike oceanic spray, wind-blown dust tends to dominate inland continental areas and becomes increasingly subordinate in coastal regions. It is the main source of silica, iron and aluminium in precipitation (Sugawara, 1967), and may also contribute significant amounts of cations such as $Ca^{2+}$ and $K^+$. An early report of this input source is the work of Junge and Werby (1958) for the south-western United States. Once suspended in the atmosphere, such particulates may contribute to the neutralisation of acid pollutants to some extent. The precise nature of the dust is often not well established, but it may originate from soil (especially during dry soil cultivation), quarrying, vegetation, natural fires, etc. The amount and chemical composition of dust may therefore be expected to vary regionally as a result of changes in geology and vegetation cover and distance from point sources, and seasonally in response to cultivation and, especially, snow cover. Rutherford (1967), for example, found higher levels of calcium, nitrate and silica associated with limestone areas compared to metamorphic rock of the Canadian Shield. Thornton and Eisenreich (1982) examined the influence of land use on the composition of wet deposition along a transect across northern Minnesota. There is a steady transition from west to east from prairie–agriculture to prairie–forest to temperate coniferous forest, with a change from calcareous to non-calcareous soils in the same direction. The results suggested that soil sources were dominant in the west, but in the afforested eastern region became less important compared to anthropogenic pollutants.

*Lightning*
It has long been recognised that the high energy levels available locally
in lightning may be adequate to convert substantial amounts of molecular nitrogen to $NO_x$. Recent attempts to estimate just how significant
this factor is have been summarised by Tabatabai (1985). There is little
agreement in the studies published, which is hardly surprising bearing in
mind the variability of the energy source and of prevailing climatic fac-

Fig. 2.2. Typical variation in precipitation chemistry over the duration of a
single storm event. Filled circles = pH; filled triangles = sodium; open
triangles = calcium; filled squares = magnesium; open circles = potassium.

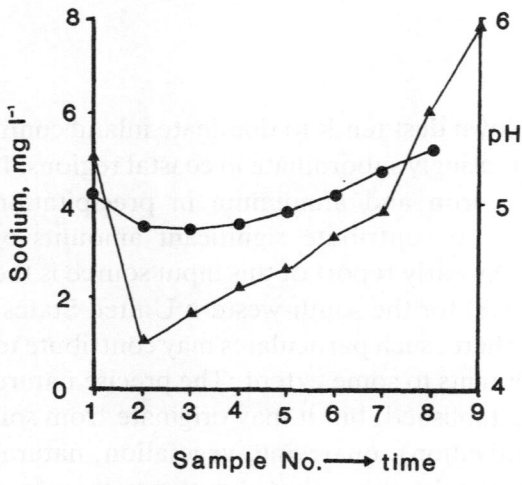

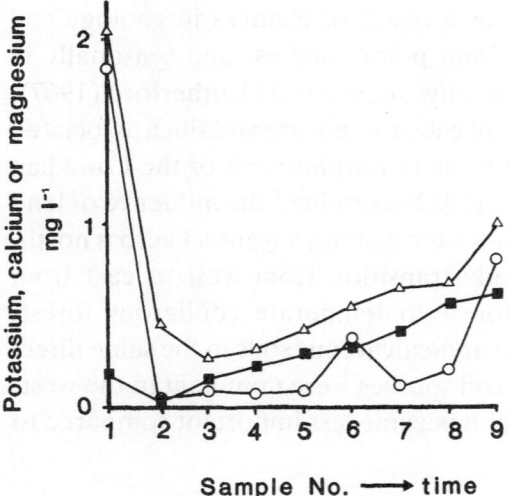

tors, and the gross difficulty of trying to base reliable estimates on highly circumstantial evidence.

## Seasonal trends in precipitation chemistry

From the preceding discussion, a number of seasonal trends in precipitation chemistry might be anticipated, depending upon seasonal climatic patterns. Some of these have been mentioned already, but it is appropriate to summarise major expected trends briefly here.

Pollutants from fossil fuel combustion might be expected to increase in winter if fuel consumption increases (Reid *et al.*, 1981). If, on the other hand, precipitation also increases in winter, ion concentrations of such pollutants may remain virtually unchanged, although total wet deposition inputs would increase. Increased storminess of the oceans in winter tends to increase maritime-derived aerosol inputs (Reid *et al.*, 1981). Snow cover may considerably alter terrestrial dust contributions and dry deposition patterns. Throughfall chemistry would clearly be modified to a different extent in deciduous forest in different seasons. Seasonal patterns in dominant wind speed and direction may influence all input sources. Seasonal cultivation trends may substantially modify the relative importance of terrestrial dust.

## Short-term trends in precipitation chemistry

The sub-cloud scavenging mentioned earlier in this chapter (washout) often results in dramatic changes in precipitation chemistry with time even during a single storm event. Some typical results we have obtained with a fractionating rain gauge are shown in Fig. 2.2 (Edwards and Cresser, 1985). It is important to bear such behaviour in mind when investigating acid precipitation effects, because, for brief periods early in a storm, exceptionally low pH values may be occurring, and the bulk precipitation sample for the entire storm may give no indication whatsoever of such an effect.

## Hydrology of upland catchments

Precipitation from previous storms maintains the world's river systems, and therefore there is obviously a strong link between precipitation quantity and river runoff. However, the proportion of incident deposited water which eventually reaches a river channel and the time delay involved depend upon a variety of interlinked factors. Both the quantity of drainage water and the speed with which precipitation is transferred from the soil surface to a stream or lake often show marked seasonal effects, transfer being more rapid and complete in cooler and/ or wetter seasons. A characteristic typical of the upland areas discussed earlier is the possibility of major contributions to annual precipitation totals resulting from snow-falls, with associated low air temperatures, often resulting in soil temperatures below 0 °C for a variable period. Apart from the problems of quantifying snowfall on exposed, wind-swept sites, the build up of a snowpack as a temporary water storage reservoir and the occurrence of less permeable frozen soils have important effects on runoff patterns to streams and lakes. Fig. 2.3 illustrates the possible major mechanisms and routes involved in the generation of stream runoff for upland areas.

The complex nature of stream runoff generation, and its dependence on seasonal climatic influences is currently an area of tremendous interest, which is receiving significant research inputs from a range of scientific disciplines. The relative importance of atmospheric, vegetational, geographical and human factors and the interactions between them are discussed in this book where they are relevant to water acidification.

Fig. 2.3. Major hydrological pathways in upland catchments. 1 = overland flow; 2 = return flow; 3 = throughflow; 4 = drainage to water-table.

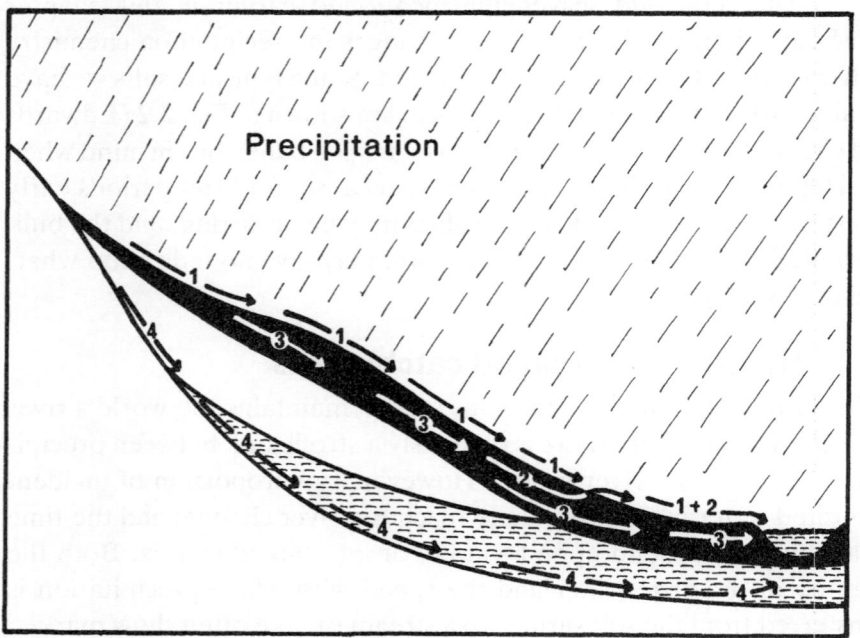

The response of any particular stream to rainwater falling upon its drainage basin is critically dependent upon the rainfall duration (McCaig, 1983) and intensity, and upon the antecedent moisture conditions (Weyman, 1975; Whipkey and Kirkby, 1978). It is also influenced by the soil's infiltration capacity (Dunne, 1978), hydraulic conductivity and thickness (Kirkby and Chorley, 1967), and hillslope form (Anderson and Burt, 1978) and steepness. Superimposed upon these factors is the impact of vegetation cover. The type of vegetation is important both because of changes in the extent of interception (i.e. its effect upon the amount of water reaching the soil surface) and also, especially in the long term, because of its influence on soil profile development and drainage characteristics. Vegetation type may also exert a direct influence upon precipitation chemistry, quite apart from its modifying effect upon water flux and movement through soil.

*Response of streams or lakes to storms in upland catchments*
Upland catchments in areas which presently or in the future may experience periods of low water pH are generally characterised by steep slopes, often shallow mineral soil horizons under organic-rich surface horizons, sometimes with rock outcrops, and moderate to low soil and air temperatures (and hence modest evapotranspiration). When it rains, water falling upon slopes wets up the soil and eventually penetrates to the underlying rock surface or to any impermeable horizon present. If horizons become saturated, water then flows laterally downslope to the stream or lake. Where natural pipe structures exist, very rapid drainage downslope may occur even before the soil is fully saturated (McCaig, 1983). Generally, rapid lateral flow occurs during heavy rain within very few hours of the start of a storm. The water residence time in the soil is short (hours rather than days), especially in smaller catchments, and stream discharge rises rapidly. Typical behaviour for an upland stream we have studied in north-east Scotland is shown in Fig. 2.4. Because of the episodic nature of precipitation events and the often limited water storage capacity (thin soils), it follows that such upland streams spend the majority of the total time at or near a baseflow condition. This characteristic behaviour is reflected in the typical annual flow duration curves of Fig. 2.5, which show the percentage of total time a specified discharge is equalled or exceeded in each of two typical streams. As will be seen later in this chapter, stream-water chemistry can change dramatically with time during storm episodes, and it is therefore very important in studies of water acidification in upland

catchments to make sure that the sampling program is adequate and includes intensive stream sampling during and immediately after storm events (see Chapter 4). Even regular weekly sampling may only 'catch' one or two high discharge events in a year (Edwards *et al.*, 1984).

There has been considerable development in our understanding of the concepts behind stream hydrograph responses to individual storm events. While it would be inappropriate to discuss the topic further in the present context, interested readers should look at papers such as those by Langbein and Iseri (1960), Amerman (1965), Hewlett and Hibbert (1967), Kirkby and Chorley (1967) and Jones (1979), or at a book such as *Hillslope Hydrology* (Kirkby, 1978), or the review by Linsley (1967).

Complications in the general discussion of upland catchment hydrology arise because areas of differing hydraulic conductivity within the soil profile have not often been considered. The importance to water acidification of features such as relatively impermeable iron pans or indurated horizons or reduced permeability at depth due to overburden is their potential for allowing bypassing of soil mineral horizons. They thus may increase the importance of stormflow from upper acidic organic soil horizons. These perched water-tables may develop wherever a sequence of soil horizons occurs with different hydraulic

Fig. 2.4. Typical relationship between rainfall and river discharge in an upland catchment in north-east Scotland.

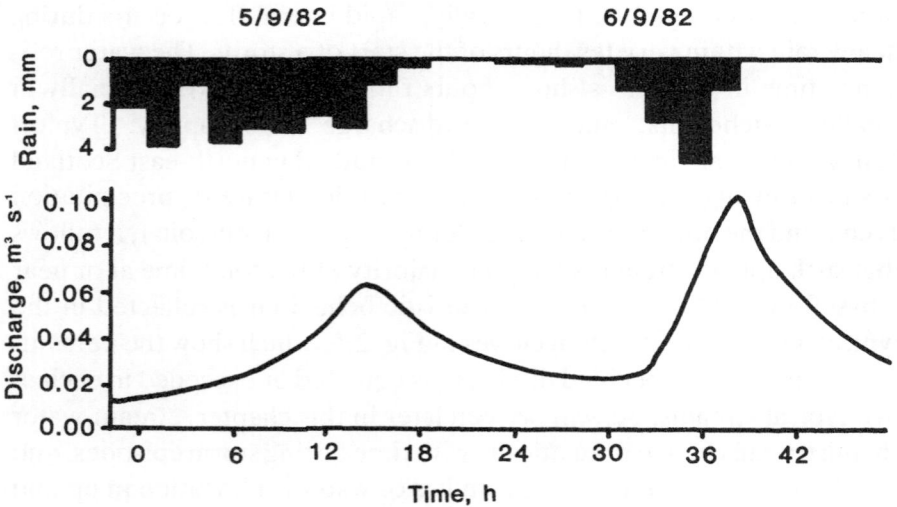

conductivities (Childs, 1957). Lowery *et al.* (1982) have attributed observations of overland flow from slopes of the Willametter Valley (Oregon) to be due to ephemeral perched water-tables and high antecedent moisture.

The changes in river-water chemistry with time associated with storm events reflect changes in the relative contributions to total discharge of water following different hydrological pathways. Usually water flowing laterally downslope through surface organic horizons makes a greater contribution during storms, with a subsequent reduction in the relative contribution of water draining from underlying mineral horizons. When rainfall is particularly heavy, precipitation may exceed the soil infiltration capacity, resulting in overland flow. Velocities of overland flow are appreciably lower than open channel flow velocities. This is an important point to bear in mind, because the water has time to equilibrate chemically with the soil surface.

The importance of pathways taken by percolating water through soil to each horizon cannot be stressed enough. Ugolini *et al.* (1977) and Dawson *et al.* (1978), studying soil solution collected at various stages of its journey through a fir forest and podzol soil in the Central Cascades,

Fig. 2.5. Typical flow duration curves for two upland catchments in north-east Scotland.

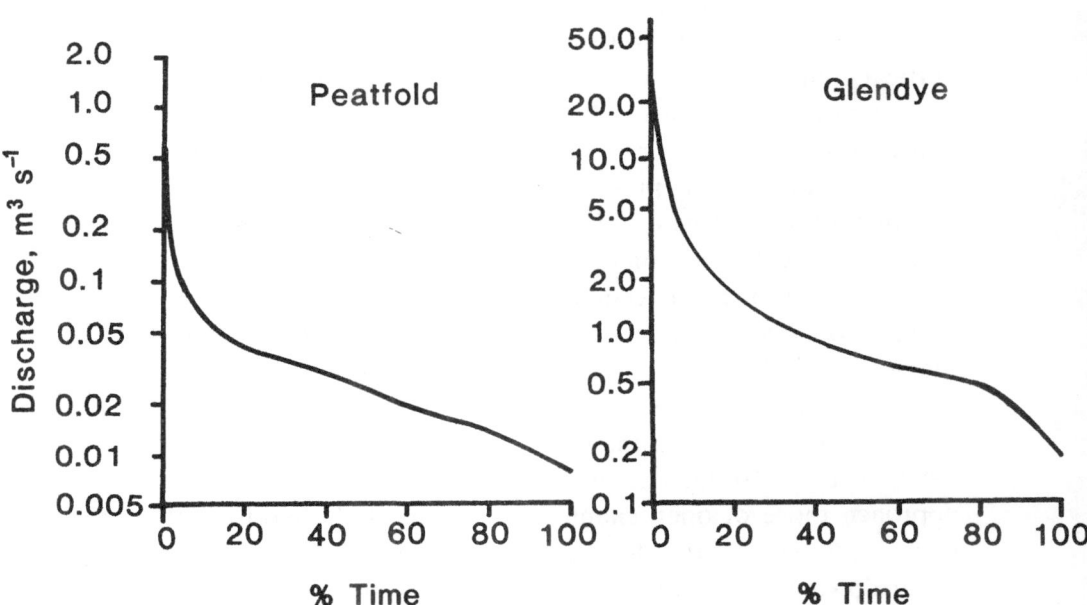

Washington, highlighted the fact that reaction products at one depth are the reactants for the next depth, a very important concept.

## Mechanisms of water transport through soil

On entering the soil, rainwater may be stored or move towards the stream channel, depending upon the antecedent moisture conditions. The hydraulic conductivity of a soil, which reflects the speed at which the water front moves down through a soil profile, depends upon pore geometry and continuity, and obviously is influenced by the soil texture (particle size distribution) and antecedent soil moisture content. For a soil with a low moisture content only the smaller pores would be expected to be filled, and these transmit water much less readily than the larger pores. Infiltrating water varies the moisture content of the surface soil until the hydraulic conductivity and rainfall intensity are equal, resulting in a downward-percolating wetting front, followed by a long period of redistribution of water in the soil profile. Soil is always in a state of flux with respect to its moisture status; evaporation, transpiration and drainage removing water, while precipitation tops up the system. A continuous range of pore diameters may exist in soil, but for convenience soil moisture movement may be divided into two component parts, matrix flow and pipe flow (Atkinson, 1978). This division is somewhat arbitrary and is based on the particular conducting void diameters and potential flow velocities.

Matrix flow is defined as flow through inter-granular pores and smaller structural voids. Matrix flow may be divided into downslope and vertical components and may occur under both saturated and unsaturated conditions (Atkinson, 1978). Water movement occurs in response to a hydraulic potential gradient, which arises from the combined effects of gravitational head, pressure (capillary) potential and the osmotic potential of the soil water (Hillel, 1971). Water moving in this way may be expected to make intimate contact with mineral surfaces and to have undergone considerable chemical modification.

Pipe flow occurs through large voids (pipes), 1 to 2 cm or more in diameter, which form open passageways through the soil. Saturation of the soil around the pipe is regarded as an essential prerequisite to the initiation of pipe flow at the reported velocities (Gilman and Newson, 1980), which may far exceed overland flow velocities and may even approach those of open channel flow. Pipe flow therefore offers the most rapid mechanism for the transmission of subsurface drainage water down a slope (McCaig, 1983). Such water may have made relatively little contact with mineral soil if flow is restricted to the peaty surface layers of upland podzols (Atkinson, 1978). Other potential regions susceptible to pipe flow include zones above impeding layers such as iron pans and those within or just below horizons of greater aggregate stability (Jones, 1979). Considerable amounts of water can only move through large interconnected pores such as worm or root channels or pipes when the soil is at or close to saturation (Mosley, 1982). Therefore, it might be expected that pipe flow will only make a significant contribution to discharge during prolonged moderate to heavy rain when upper horizons become saturated. This appears to be the case in practice. Water originating from matrix flow mainly through mineral horizons has a relatively much longer residence time and may be expected to dominate low or base flow.

The soil moisture content will vary, depending on the profile's position within the catchment area. Relatively higher moisture contents at a particular depth occur for soils on lower slopes, resulting in increased hydraulic conductivity (a function of moisture content) downslope. It therefore follows that a smaller quantity of infiltrating water will be required in the topsoil on lower slopes to raise the conductivity of the soil sufficient to transmit water. As a result, the relative contributions to storm stream flow from particular areas will vary with time. Before we can assess the significance of the facets of soil water movement through soils discussed above to water acidification, we need first to consider the nature of soil/water interactions.

## Soil/water interactions

Soil is best regarded as the layer of the lithosphere which is currently undergoing chemical, physical and biological change. Thus soil generally lies above its parent material, the unaltering mineral matrix. The parent material may itself have been altered *in situ* at some stage, in a previous phase of weathering or soil formation. It may be solid rock or redeposited, modified rock (glacial till, for example), which has been transported by physical forces such as water or glacier movement, wind, or gravitational force in combination with other disturbing forces. Underlying bedrock may not be related to the soil parent material. Thus water which passes through soil interacts with a chemically, physically and biologically dynamic system. This has two important consequences as far as freshwater acidification is concerned. The first, and perhaps the most obvious, is that it is necessary to consider precisely what chemical, physical and biological processes occur when

water and soil interact and how they affect water acidity. The second is
that we must consider what governs the present state of the soil and its
capacity to interact in different ways. Many soil properties are generally
fairly well buffered, natural changes other than seasonal trends under a
stable climatic regime occurring over decades or centuries rather than
weeks. It is particularly appropriate to consider the nature of the
relevant processes involved in interaction of soil solid-phase components with water, because these also govern the present soil chemical
properties. Three concepts are particularly important, mineral weathering, ion exchange and adsorption–desorption, and biological activity,
and each is considered in turn.

*Mineral weathering*
Silicate mineral weathering proceeds via hydrolysis, i.e. the replacement of one or more cationic species from a mineral matrix by hydrogen ions. The latter may originate from carbonic acid, from organic
acids of biological (including microbiological) origins or from wet or dry
deposited acid components from the atmosphere. The latter sources are
generally accepted to be of little consequence relative to carbonic acid
in the absence of human activities. Unpolluted rain equilibrates with
gases in the atmosphere through which it falls, and would thus have a pH
somewhat below 5.6, the pH of distilled water equilibrated with
'normal' air. The soil atmosphere may be enriched 100-fold or more in
carbon dioxide relative to above-ground air, and thus carbonic acid
originating from respiration plays a major role in mineral weathering.
Uptake of base cations or ammonium ions by plant roots results in
release of an equivalent amount of hydrogen ions at the root/soil interface, although in the absence of cropping this must be regarded as a
cyclic rather than a one-way process.

   Mineral dissolution reactions may be congruent, leaving no solid
residue, e.g.

$$Mg_2SiO_4 + 4H_2CO_3 \rightleftharpoons 2Mg^{2+} + 4HCO_3^- + H_4SiO_4 \qquad (2.1)$$
olivine (forsterite)

Alternatively they may be incongruent, in which case a solid residue
remains, e.g.

$$4KAlSi_3O_8 + 4H_2CO_3 + 18H_2O$$
potassium feldspar
$$\rightleftharpoons 4K^+ + 4HCO_3^- + Al_4Si_4O_{10}(OH)_8 + 8H_4SiO_4 \qquad (2.2)$$
kaolinite

Mineral dissolution rate in contacting water is often initially rapid with a subsequent decline. Causes for the reduction in rates include the increase in soluble components in the reacting solution as the weathering reaction proceeds, or sometimes the formation of surface coatings of secondary mineral materials, with the dissolution rate controlled by diffusion of cations released by weathering through the coating (see e.g. Wilson, 1975; Henderson, 1982). If water percolates slowly downwards through a column of crushed rock over a long period, the descending water becomes progressively more saturated with respect to the most weatherable mineral present. Weathering is therefore most advanced for the most weatherable mineral close to the surface, and negligible at depth. This trend is what is found in practice in some soils (see e.g. Wilson, 1967). The most readily weatherable mineral becomes progressively more abundant with depth; less readily weatherable minerals therefore becoming apparently less concentrated. The decline is of course relative.

The primary minerals were originally formed under much higher pressures and temperatures and are thermodynamically unstable at the conditions prevailing today, resulting in weathering and the formation of more-stable secondary minerals. In general, the relative resistance of minerals to chemical breakdown has been shown (e.g. Goldich, 1938) to follow closely Bowen's reaction series with local variations due to both physical (grain size and microcrack structure) and chemical (composition) variations.

Although it would be inappropriate to consider in detail here the quantitative mineral weathering behaviour of a wide selection of minerals, it is worth considering the extent to which weathering rate is likely to be influenced by factors such as soil pH and soluble salts present. Muscovite is a useful example to work with to indicate the hypothetical principles involved. It may weather according to the equation:

$$KAl_2(AlSi_3O_{10})(OH)_2 + 10H^+ \rightleftharpoons K^+ + 3Al^{3+} + 3H_4SiO_4 \quad (2.3)$$

The log of the equilibrium constant for this reaction is 13.44 (Lindsay, 1979). Thus:

$$\frac{(K^+)(Al^{3+})^3(H_4SiO_4)^3}{(H^+)^{10}} = 10^{13.44} \quad (2.4)$$

$$\log(K^+) + 3\log(Al^{3+}) + 3\log(H_4SiO_4) + 10\,pH = 13.44 \quad (2.5)$$

In the absence of any source of $K^+$, $Al^{3+}$ and $H_4SiO_4$ apart from muscovite,

$$3(K^+) = (Al^{3+}) = (H_4SiO_4) \tag{2.6}$$
$$\therefore \log(K^+) + 6\log3 + 6\log(K^+) + 10\,pH = 13.44 \tag{2.7}$$
$$\log(K^+) = 1.51 - 1.43\,pH \tag{2.8}$$

For a pH-buffered system such as a soil, it follows that muscovite dissolution could be expected to increase *ca* fourteen-fold for each unit fall in soil pH, as the plot of $\log(K^+)$ vs pH in Fig. 2.6 shows. In practice, $K^+$ inputs from other sources such as atmospheric inputs and inputs from plant litter decomposition and cation exchange sites give $K^+$ concentrations well above the muscovite equilibrium concentration above pH 5, and weathering would therefore be depressed. In more acid horizons, the additional elements released as a result of accelerated weathering in acidifying upper mineral horizons are not necessarily leached from a soil profile. As water infiltrates the soil to greater depth, pH tends to increase and many species will be retained as a consequence

Fig. 2.6. Theoretical relationship between $K^+$ and pH for muscovite in distilled water.

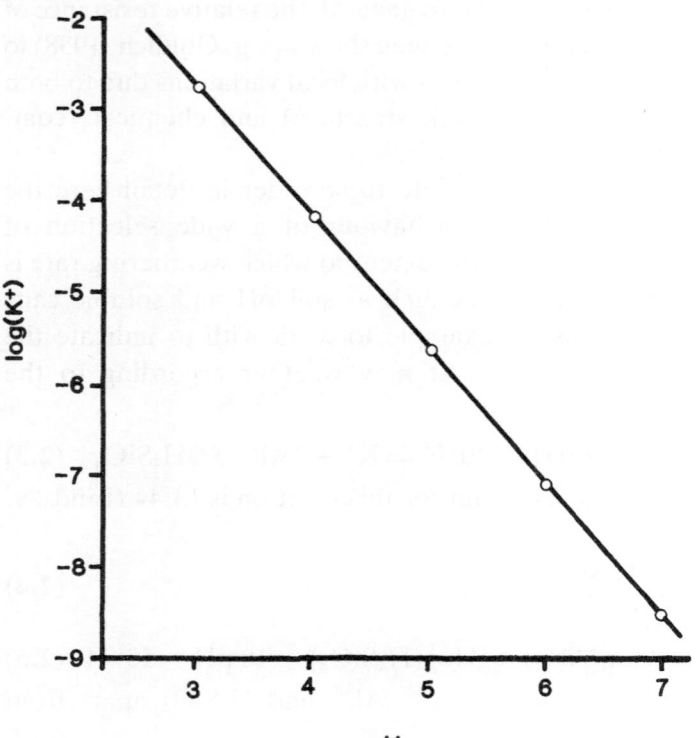

of secondary mineral formation and/or recrystallisation. Moreover, the cations released by weathering will participate in cational exchange reactions at organic matter or clay mineral surfaces, which again limits the leaching loss.

To obtain a quantitative indication of the effect of $K^+$ inputs upon muscovite weathering, it is best to rearrange equation (2.5) as follows:

$$\log(Al^{3+}) = 4.48 - 0.33 \log(K^+) - 3.33 \, pH - \log(H_4SiO_4) \quad \textbf{(2.9)}$$

It is then possible to plot graphs of $\log(Al^{3+})$ vs $\log(H_4SiO_4)$ at pre-selected pH and $(K^+)$ values. This has been done for $10^{-3}$, $10^{-4}$ and $10^{-5}M$ $(K^+)$ in Fig. 2.7 at pH 4 and at pH 7. If muscovite is the only source of $Al^{3+}$ and $H_4SiO_4$, then increasing the $K^+$ activity 100-fold from $10^{-5}M$ reduced the equilibrium $\log(Al^{3+})$ and $\log(H_4SiO_4)$ values from $ca$ $-3.6$ to $ca$ $-3.0$; in other words $(Al^{3+})$ and $(H_4SiO_4)$ are approximately halved. Thus pH effects, because of the relatively high number of protons generally involved in the weathering reaction, tend to be more important than other common ion effects. The relatively high $(Al^{3+})$ values predicted at pH 4 in Fig. 2.7 would not be expected in practice, since gibbsite should precipitate at this pH when $\log(Al^{3+})$ $\approx -3.8$ (Lindsay, 1979). Such processes produce the sort of coatings discussed earlier which may limit weathering rates in practice.

It is appropriate to consider at this point what happens when a dilute acid solution interacts with a mineral assemblage. Assuming that no additional acid input is available, hydrolysis will commence according to the appropriate mineral weathering equation. In the process, $H^+$ is neutralised and hydrolysis slows down. Eventually, if contact time is long enough, an equilibrium is reached at which point ionic activities in solution are correct for each species at the equilibrium pH. If however the acid is replenished, as happens for example when carbon dioxide is introduced from microbial or root respiration, further weathering will occur until the mineral equilibrium appropriate to the pH of the water/carbon dioxide system is eventually attained. In soil, the pH of the buffered soil/water system would be the crucial factor. As will be seen later, water draining from such mineral soils often has a pH close to or above 7, not just because of partial or complete acid neutralisation by weathering, but also as a result of outgassing of dissolved carbon dioxide. It should be remembered, however, that the kinetics of mineral weathering are often slow, so that thermodynamic equilibrium may often not in fact be reached in a field situation. Furthermore, the overall process is complicated by the involvement of ion exchange, redox and adsorption–desorption reactions.

Areas which are generally thought to be susceptible to soil and water acidification tend to be those where the outcropping rock and soil parent materials are naturally acidic (i.e. low in base cations) in nature. Thus areas with soils evolved from granite and other acid igneous rocks, most metasediments, grits, quartz sandstones and decalcified sandstones and some Quaternary sands and drift materials are thought to be particularly high risk areas (Kinniburgh and Edmunds, 1984). Risk is less where the parent material has a higher content of readily weatherable minerals with their greater associated neutralising capacity. Thin soils are particularly at risk, simply because the thinner the soil the greater the possibility of the most readily weathering

Fig. 2.7. Theoretical effect of $K^+$ upon $Al^{3+}$ in solution from muscovite equilibrated with $10^{-3}$, $10^{-4}$ and $10^{-5}$M $K^+$ solutions at pH 4 and pH 7.

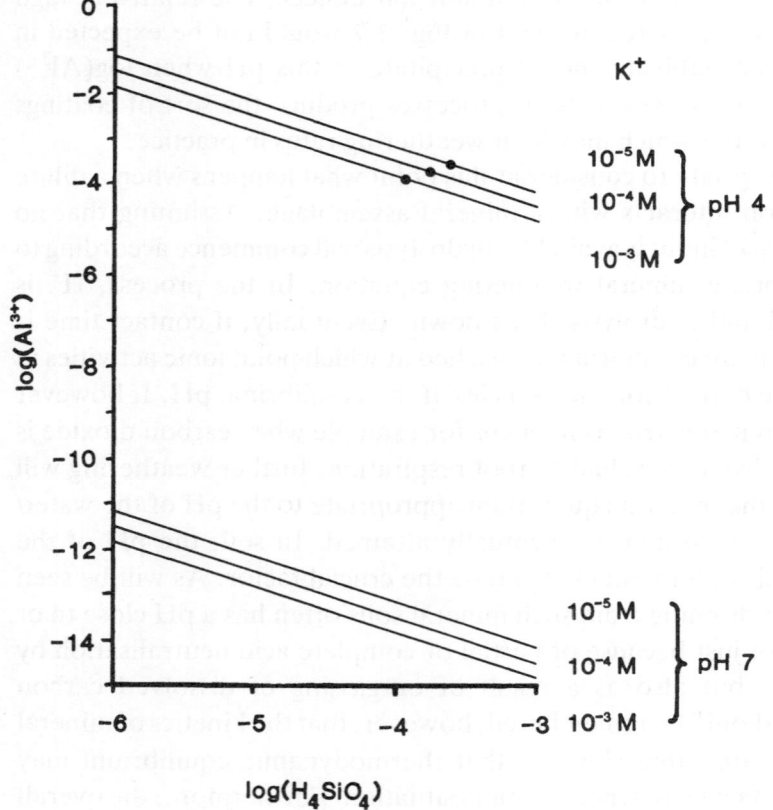

minerals being exhausted. Generally, too, the higher the precipitation, the greater the weathering-derived output, and the greater the rate of weathering of such minerals. These important points will be discussed again later when the significance of drainage basin characteristics is discussed.

*Cation exchange in soils*
One of the most important physico-chemical properties of soils is their ability to participate in ion exchange reactions. Cation exchange occurs at negatively charged surface exchange sites on soil organic matter or clay minerals. The former are primarily carboxylic acid or phenolic OH groups. The extent of dissociation of these groups, and hence the cation exchange capacity of soil organic matter, is pH dependent. The negative charge of clay minerals arises as a result of clay mineral lattice edge effects or because of isomorphous substitution (i.e. displacement of one lattice cation by another of similar size) by ions with lower positive charge. Cation exchange reactions in soils are very rapid, much more so than mineral weathering reactions. Thus cation exchange reactions are responsible for the short-term buffering of drainage water chemistry even when residence time of water in the soil is very short. Geochemical weathering, on the other hand, is primarily responsible for long-term buffering of drainage water chemistry.

For an adequate overall concept of acidification of freshwaters, a detailed knowledge of soil reactions is required. The ion exchange reactions discussed in this section are vitally important. Adsorption–desorption exchange reactions between ions in soil solution and the solid mineral and organic phases result in delay in solute species movement through the soil profile. Rates of ion transport are therefore generally much slower than estimates of water flux might suggest to an extent which depends upon a variety of factors including pH, concentration, and relative specificity of adsorption–desorption reactions. Soils may be thought of as an example of a natural ion exchange chromatographic column.

Consider what happens during a precipitation event. As water wets up and infiltrates the soil, existing soil water solute is diluted and new solute (from the precipitation) is introduced. The resulting solution equilibrates rapidly with the cations on the exchange complex, the composition of the resulting drainage water depending upon the compositions of both the exchangeable cation mixture and the initial interacting water chemistry. Some of the cations displaced from exchange sites are leached out of the soil profile with drainage water along with associated anions. Some of the incident cations may be retained. Unless the soil is wholly organic, weathering will also contribute to the drainage water solute chemistry to some extent. If the rainfall is particularly heavy, water may flow overland, simply because it cannot infiltrate the soil fast enough. Such water interacts with surface soil exchange sites. When the rain stops, water drains until the soil approaches the state known as field capacity. Lateral flow on lower slopes of upland catchments may continue for some time. In the longer intervals between storm events, geochemical weathering proceeds, replenishing all or some of the cations leached from exchange sites during earlier events. This will be assisted by elevated carbon dioxide building up in the soil atmosphere. Initially, $H^+$ ions from exchange sites react with the weathering minerals to liberate base cations, which in turn may be adsorbed onto exchange sites. Provided time allows, the final soil solution solute chemistry will satisfy the conditions imposed by soil atmosphere $CO_2$ equilibrium, ion exchange equilibria and geochemical weathering equilibria. This situation is, however, even further complicated by cation uptake by plants, itself effectively an ion exchange process, and possible upwards movement of soil water as a result of evapotranspiration.

When soil weathering and atmospheric inputs produce sufficient base cations, and ion exchange equilibria are such that the base saturation of the exchange complex remains high, the soil pH remains relatively stable. When, however, geochemical weathering is incapable of adequately replenishing base cations leached from exchange sites during or immediately after storms, soil pH and base saturation start to fall. The problem is exacerbated by high rainfall, by cropping, or by a naturally acidic nature of the parent material of the soil, as mentioned earlier.

A comparison we have made of two catchments with different parent materials, but similar climate, topography and land use, serves to illustrate the approximate long-term time scales of serious soil acidification (Edwards et al., 1985). Where the parent material was granite, the soils, particularly on the upper slopes, were already approaching the end of an 11 000-year post-glacial weathering period where leached calcium, relatively the most mobile element, could be readily replaced by mineral weathering. For the more basic quartz–biotite–norite parent material, reserves should suffice for a further 12 000 years under the present land use regime. Although a number of simplifying assumptions had to be made in arriving at these estimates, they do serve to show why catchments with soils derived from acid igneous rocks such as granite are those at risk from soil and water acidification.

*Anion exchange reactions*

Anion exchange in the soils of interest in the present context is primarily associated with hydrated iron and aluminium oxides. Anion exchange capacity is more sensitive than cation exchange capacity to changes in soil solution pH and electrolyte concentration. Moreover, the selectivity to adsorption of different anions is much larger than the selectivity of cation exchange sites to diverse cations. Chloride and nitrate are non-specifically adsorbed onto positive soil surfaces, while the remaining major anions, including organic matter, are specifically adsorbed onto individual sites forming more stable chemical bonds (ligand exchange), resulting in a greater affinity for soil surface than would be expected from their concentration in solution. This leads to a marked order of preference:

$$SiO_4^{3-} > SO_4^{2-} > NO_3^- > Cl^-$$

The high iron and aluminium oxide concentrations in the B horizons of podzolised profiles are therefore capable of adsorbing considerable amounts of sulphate (where this is the mobile anion). In effect $SO_4^{2-}$ is adsorbed and $OH^-$ displaced into solution. Since the soil pH is invariably buffered, the $OH^-$ reacts with $H^+$ from cation exchange sites, and base cations ($Ca^{2+}$, $Mg^{2+}$, $K^+$, $Na^+$) are exchanged in their place in equivalent amount. Thus the process of anion adsorption leads to retention of base cations, and limits the rate of soil acidification. As will be seen later, a possible adverse effect of high sulphate pollution is sulphate saturation and accelerated base cation leaching and soil acidification.

## Relevant soil formation processes

Drainage water pH depends upon the pH of the soil horizon through which the water last passed. Therefore a question that must be asked at this stage is: Just how acid could soils become in the total absence of the anthropogenic influences which will be discussed in Chapter 3? The quantitative answer unfortunately is not clear-cut, because of the absence of suitable control sites allowing direct measurement in any given area. Nor is it clear-cut just how fast (i.e. whether over one or two centuries or a longer time scale) soils may quite naturally acidify once geochemical weathering is no longer capable of offsetting leaching losses at an adequate rate and base saturation falls to a new equilibrium level.

Acidification of uncultivated soils is eventually accompanied by a marked change in the appearance of the soil profile, as well as in soil chemical properties. Fig. 2.8 is an attempt to illustrate the relevant trends in profile development with increasing acid nature of parent rock, increasing altitude, increasing time or increasing precipitation amount, alone or in any combination. A similar diagram has been used by FitzPatrick (1980) in his excellent account of soil formation processes. When base cations leached from exchange sites are replaced at a relatively high rate, base saturation and soil pH remain relatively high (pH 6–7). The soil remains fertile and microbial activity is high, as is the activity of macro-organisms. Thus the soil remains well mixed and a typical cambisol (brown earth) profile develops.

Fig. 2.8. Trends in soil profile development in a leaching climate with increasing acid nature of parent rock, time or precipitation amount. Ah = A horizon with humus accumulation; $B_s$ = B horizon with iron and aluminium oxide accumulation; O = organic horizon; E = leached horizon; Eg = E horizon with gleying; Pk = placon or iron pan.

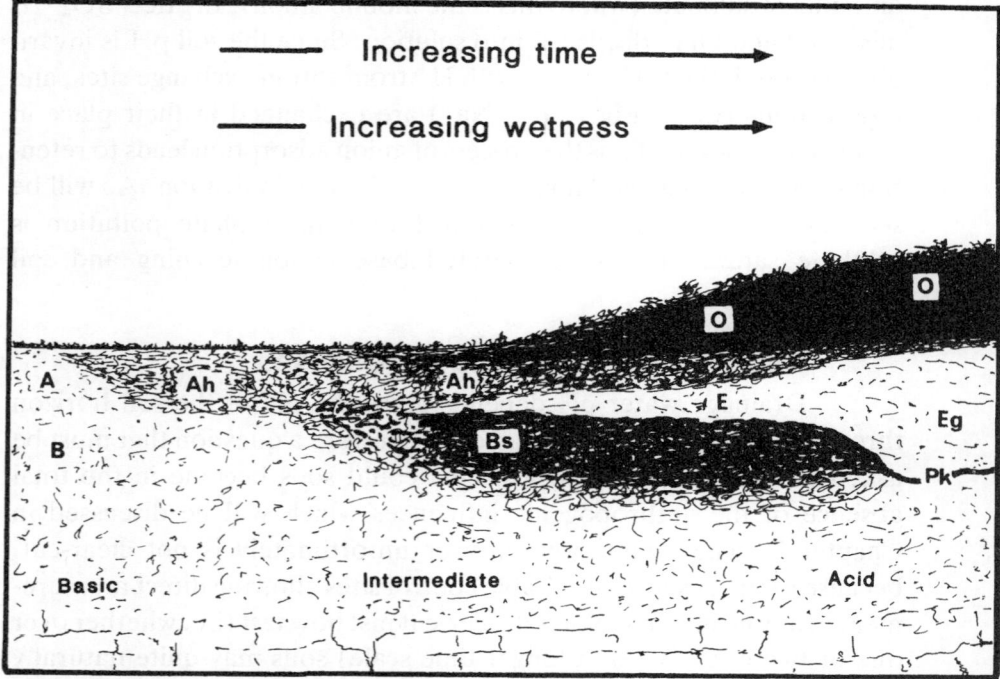

With continued advancement of weathering, the stage is eventually reached when mineral weathering becomes progressively less and less capable of replenishing leached cations. Base saturation and soil pH start to fall, particularly close to the surface. As the soil becomes more acidic (pH 5.0), activity of bacteria is depressed, along with the activity of macro-organisms, fungi becoming the dominant decomposers for plant litter. Mixing to depth is steadily reduced and organic matter starts to accumulate in surface horizons. The increasing soil acidification may itself encourage colonization by acid-tolerant plant species such as *Calluna vulgaris* or *Pinus sylvestris*. These in turn produce highly acidic litter horizons. The reduced pH at the surface increases the solubility of iron and aluminium species, which are then, partially as organic chelate complexes (see e.g. Evans, 1980), translocated down the profile until they are deposited in the higher pH, B horizon which then starts to take on an orange/brown coloration. Eluviation of iron from the surface mineral horizon leads to the formation of a leached E horizon. Gradually the extent of this process, known as podzolisation, increases, leading to the formation of a well-defined podzol.

The next stage in the development sequence is the formation of a placosol, in which case the iron is deposited in a relatively narrow, impermeable band a few millimetres thick, rather than over the 100 mm or more of the typical podzol B horizon. Depending upon climate and local topography, the formation of such an iron pan (placon) may be associated with increased hill peat formation, as may be the later stages of podzolisation. Hill peats formed under these circumstances tend to be naturally very acidic, and represent the final stage of the acidification sequence. The formation of an iron pan may have a secondary effect of modifying hydrological pathways, because continuous placons are generally highly impermeable. As discussed later, if this confines lateral water to surface acid organic and highly leached mineral (E) horizons, it may have far-reaching consequences for water acidification.

## Natural acidity of soils

It is appropriate now to return to the question asked at the beginning of this section: How acid can soils become naturally? We know at least how acid they become with the present precipitation chemistry. In the areas we and colleagues have studied in detail, it is quite clear that the pH of podzol surface horizon soils derived from granite (Glendye in the Grampian Region of north-east Scotland), lying in the range 3.4 to 3.7, may be more than a whole pH unit lower than that of incoming precipitation (Reid, 1979). At Peatfold, Glenbuchat, where the parent material is quartz–biotite–norite and geochemical weathering is still substantial in the A horizon, only on the upper slopes was soil pH significantly below precipitation mean pH; values for four podzol surface horizons were 4.2, 4.3, 4.6 and 4.1. On the middle slopes at that site, cambisols were dominant, with surface horizon pH values in the range 4.8 to 4.9, coincidentally quite close to the mean rain pH (Creasey, 1984). Both of the above sites were heather moorland. In some forested areas, lower surface horizon pH values would be anticipated.

Once conditions at a site deteriorate to the point at which hill peat accumulation becomes substantial, the situation with respect to horizon acidity may worsen considerably. At Peatfold the small patch of hill peat (*ca* 2% of the catchment area) has a mean pH of 3.4 between the surface and 50 cm depth, and even below this depth (i.e. 50–100 cm) the mean pH was 3.7 (Creasey, 1984). Hill peat at Glendye was very similar, near-surface pH values of 3.5 and 3.6 being noted, with an increase to pH 3.8 at 50–100 cm. As would be expected from the preceding discussion, hill peat is much more abundant at Glendye, covering *ca* 60% of the drainage basin surface. The similarity between the peats at the two sites suggests that underlying material may be exerting little influence, the dominant factor being climate and precipitation chemistry. Soil pH determination in both cases was done on thick water-pastes.

It is instructive to compare soil pH results and soil characteristics at Glendye with those for other granite catchments, although this must be done with care since the chemical and mineralogical composition of granites is notoriously variable. A particularly interesting comparison is that with soils in the area around Galloway on the south-west coast of Scotland, where precipitation is higher and acidification of freshwaters known to be a problem. According to Bown and Heslop (1979), the organic horizons of peaty podzols of the Dalbeattie Association Carsphairn Series are well developed to a thickness of *ca* 25 cm, and have pH (water) values lying in the range 4.0 to 4.4. A weak iron pan occurs at a depth of *ca* 12 cm in the mineral soil, and induration at *ca* 50 cm. Drainage tends to be poor above the iron pan and the soils remain wet. The parent material is primarily true granite, but with some associated less-quartz-rich rocks, and associated tills. Peat thickness tends to increase with precipitation, which rises to 2250 mm $a^{-1}$ on higher ground.

The parent material of the Ettrick Association in the same area is more variable. The Ordovician and Silurian rocks are interbedded with greywackes and shales. Silica content ranges from that of an acid igneous rock such as granite to that of an intermediate rock such as andesite. As with the Dalbeattie Association, induration is dominant below *ca* 50 cm. Peaty podzols constitute the Dod Series, and their organic horizons have a pH (water) of 3.7. In this instance a strong iron pan is developed at *ca* 17 cm depth in the mineral soil. Both series tend to merge with unclassified peats, the latter having pH values around 3.6 to 3.9, which overlap those discussed above for peat from Glendye and Peatfold, in spite of the substantial climatic difference. Cuttle (1983) reported a mean pH value of 3.5 for 11 peats from upland sites in southern Scotland.

As might be expected from the preceding discussion, iron pans (placons) are a common feature of upland soils in high rainfall areas with acidic parent materials. The Soil Survey Memoir for *Soils of the Country Round Kilmarnock*, the area to the north of that just discussed, lists six series of freely drained peaty podzols. All have iron pans, at depths between 5 and 12 cm, and all have very acidic surface soils. The mean pH of surface horizons from eight profiles was 4.1, and, for the four surface horizons with >70% organic matter, 3.8 (Mitchell and Jarvis, 1956).

In Wales, also, whether or not a placon forms from acid igneous rock depends upon the precipitation amount. On the mainland just across from Anglesey, podzols of the Bodafon Series have a surface pH of 4.1 and precipitation is similar to that in Glendye (Roberts, 1958). Further inland, towards Snowdonia, precipitation increases considerably. The mean surface pH of peaty podzol surface horizons was 4.2 (five profiles; Ball *et al.*, 1969). In this instance the parent materials were mainly acid lava and ash drift materials from rhyolitic origins. Deep peat is widespread, having a pH of 4.0–4.5. Iron pans are rarely mentioned, probably existing under deep peat or peaty gley podzols (Rudeforth *et al.*, 1984). The latter also give very acid surface horizons. The acidity of Welsh soils has recently been reviewed by Thompson and Loveland (1985) who recorded a mean pH of 4.2 (± 0.4) for 16 ferric and iron pan stagnopodzols under rough grazing.

Crampton (1967) examined soils protected under archaeological structures in South Wales. Whereas podzolisation dated back to the end of the Iron Age, discontinuous iron pan formation was a Medieval feature in that area. This point could be of great importance, because we believe that continuous impermeable placon formation may lead to substantial modification of hydrological pathways and thus to serious water acidification over a time scale of one to two-hundred years. This topic, and its possible relationship to forest clearance, are considered again in Chapter 3. Crampton's work shows that upland soils eventually became sufficiently acid for placon formation, even in the absence of significant acidifying pollutants of anthropogenic origins (Crampton, 1967). This suggests that soils would naturally tend eventually towards pH values of 4.0–4.1. It is well known, too, that peat formation often occurred prior to heavy pollution.

A question mark remains over whether the acidity of peats is lowered as a result of pollution effects. The precipitation on the west coast of Scotland tends to have a higher pH than that on the east coast, and hill peats on the east tend to have slightly lower pH. Quantitative interpretation is difficult because of the higher precipitation on the west side. Leaching experiments certainly suggest that precipitation acidity may lower the surface pH of organic soils by up to a fraction of a pH unit, but it is not yet possible to be more precise than this. These experiments are discussed briefly in Chapter 3. The pH values of surface mineral horizons from four podzols evolved from greywackes in the Southern Island of New Zealand, from areas with 1800 to 2500 mm annual precipitation but in an unpolluted area, were 3.8, 3.7, 3.5 and 4.5 (Lee *et al.*, 1985), suggesting that the anthropogenic acidity pH shift will be small except when pollution is very severe.

## The importance of vegetation

Vegetation type has long been recognised as an important factor in soil development, controlling to various degrees nutrient cycling, storage, water regime and humus type. Indeed soil type may change with predominant plant species. For example, Jarvis and Duncan (1976) suggested that, along a transect where no other visible controlling factors could be seen, a change from heather to a bracken-dominated system could be accompanied by a change in soil type from a podzol to a cambisol, probably as a result of changing humus type. Accelerated rates of podzolisation have been noted in relation to tree species too. The effect is generally thought to be related to changes in type and quantity of organic compounds leached from leaves and roots. Crampton (1982) showed the development of circular zones of podzolisation around the trunks of Douglas fir and western hemlock. He suggested that very acidic stemflow could be an important contributing factor.

De Kimpe and Martel (1976) investigated the importance of vegetation type (coniferous and deciduous) on the distribution of C and Fe and Al oxides in B horizons of ten podzolic soils of the northern Appalachian Mountains. They suggested changes in the humic acid/fulvic acid ratio (higher under deciduous) occurred, with conifers producing relatively more low molecular weight substances. This was suggested to be important in profile development, especially mobilisation of Fe and Al.

Williams *et al.* (1978) compared nutrient contents, acidity and exchangeable cations in the upper 300 mm of peat beneath lodgepole pine and contiguous unplanted areas at each of six sites. N, P and K contents varied between peat types but were not significantly altered by the presence of trees. However, differences in exchangeable $Ca^{2+}$, $Mg^{2+}$ and $K^+$ were observed as a result of planting. Other similar studies have also been published.

An obvious effect of vegetation growth on river water is a reduction in the amount of water leaving a catchment as a result of increased interception, transpiration and evaporation. Each of these effects is species-dependent. Thus different vegetation types may influence drainage water chemistry via their effects upon dominant hydrological pathways, quite apart from the effects of different amounts and types of litter, different collection efficiencies for dry deposition, direct precipitation/vegetation interactions and nutrient uptakes. Quantities of elements stored in the biomass may be very significant in terms of element cycling. At Hubbard Brook, New Hampshire, for example, annual storage of elements within the biomass of an aggrading hardwood forest accounted for up to 50% of weathering products (Likens *et al.*, 1977). This may be a major feature of forested ecosystems until the forest reaches maturity (Ovington, 1965), leading to considerable pH decline in stressed ecosystems. This aspect is discussed in more detail in Chapter 3, where the significance of changes in land use is discussed in more detail.

## pH buffering of freshwaters

One of the remarkable features of many streams draining upland catchments is that, in spite of the frequently very acidic nature of the soils through which the water has drained, river-water pH is often close to 7, or at least well above soil pH for much of the year. Except where base flow is dominated by water draining from more acidic hill peats, upland stream pH falls as discharge increases during storm events and snowmelt, returning more slowly to its baseflow value only as discharge recedes. The net effect is the pH hysteresis shown in Fig. 2.9.

*The importance of carbon dioxide*
Water entering the soil equilibrates with the soil atmosphere, leading to a carbonic acid solution, the concentration of which depends upon soil pH and the partial pressure of $CO_2$ in the soil atmosphere:

$$CO_2 + H_2O \rightleftharpoons H_2CO_3 \qquad\qquad (2.10)$$

$$H_2CO_3 \rightleftharpoons H^+ + HCO_3^- \qquad\qquad (2.11)$$

For equation (2.10), the equilibrium constant, $K_1$, is given by:

$$K_1 = \frac{[H_2CO_3]}{[H_2O][CO_2]} = \frac{[H_2CO_3]}{[CO_2]} \qquad\qquad (2.12)$$

Fig. 2.9. Typical pH hysteresis effect for a storm in an upland catchment.

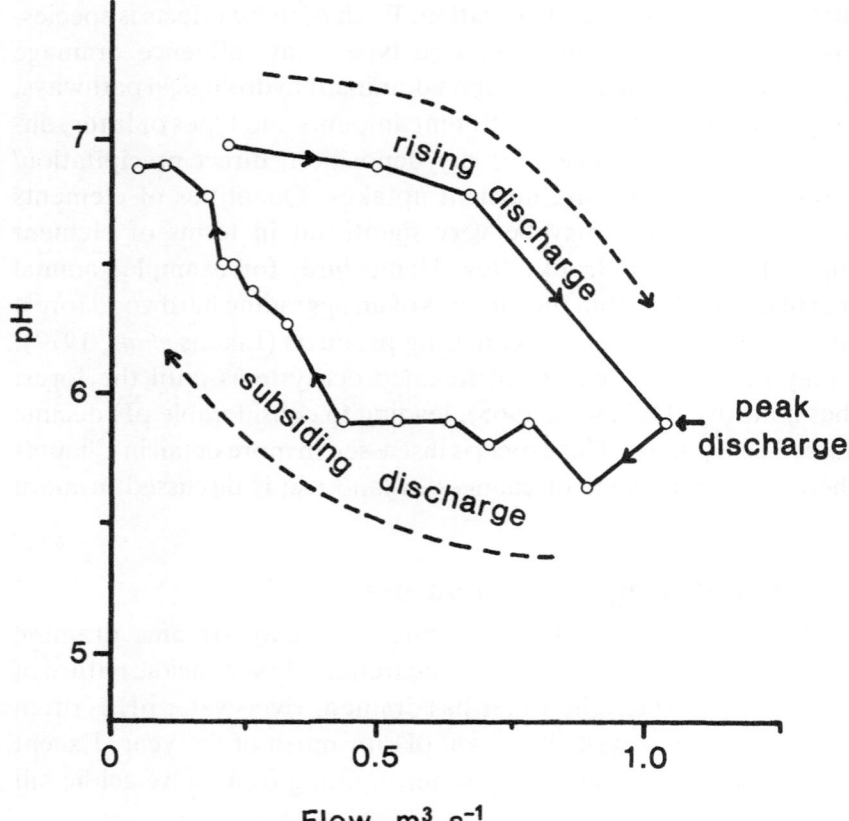

For equation (2.11):

$$K_2 = \frac{[H^+][HCO_3^+]}{[H_2CO_3]} = \frac{[H^+]}{K_1} \frac{[HCO_3^-]}{[CO_2]} \qquad (2.13)$$

Thus:

$$K_3 = \frac{[H^+][HCO_3^-]}{[CO_2]} \qquad (2.14)$$

Suppose by way of an example $[CO_2]$ in the soil atmosphere is enriched 100-fold compared to that in the above-ground atmosphere. When such water drains out of the soil, $[CO_2]$ falls 100-fold, and to maintain $K_3$, the product of $[H^+]$ and $[HCO_3^-]$ must also fall 100-fold. If this happens simply by equations (2.11) and (2.10) progressing from right to left, $[H^+]$ must decrease ten-fold, and pH will rise by one unit. If on the other hand the carbonic acid has produced bicarbonates as a result of mineral weathering, $[HCO_3^-]$ would fall by less than ten-fold, and $[H^+]$ by more than ten-fold. Thus it is quite reasonable when the $CO_2$ effect is significant to find river water with a pH value two units above that of the soil through which the water has just drained.

The existence of this effect can be very simply demonstrated by allowing river water to equilibrate with air enriched in increasing amounts of $CO_2$. This was done, for example, for water from Glendye equilibrated with the required atmospheres from suitable mixtures in gas cylinders to produce the results shown in Fig. 2.10. A useful account of the importance of the $CO_2$ effect has been published by Norton and Henriksen (1983).

The fall in pH of river water during a storm event in catchments such as those we have studied is thus due to three interacting factors:

   (i) As the storm progresses, lateral flow through surface organic horizons increases in relative importance. The water in these soil horizons has a lower pH when the soils themselves are more acidic and a lower concentration of base cation bicarbonates, so the $CO_2$ buffering effect is relatively smaller.

  (ii) The soil atmosphere $CO_2$ effect also decreases with time as $CO_2$ is removed in solution in the drainage water.

 (iii) Soluble organic acids from the surface horizons may become more significant.

The importance of $CO_2$ buffering declines when winter temperatures are low for prolonged periods. However, such seasonal trends may be masked by the influence of frozen surface horizons and snowmelt when meltwater flows over frozen soil horizons, or by heavy autumn rains.

## The significance of readily weatherable mineral depletion

From the preceding discussions it should be apparent that, for two catchments with similar climate, land use and topography, but base-rich and base-poor parent materials, the base-poor catchment will give more acidic soils. As Fig. 2.8 shows, well-developed podzol profiles are also more likely, and acidic surface organic horizons are likely to be thicker. In a high rainfall area, iron pan formation is more probable, again leading to increased prospect of deep peat formation. The greater abundance of more acidic surface horizons is what dominates river-water pH during periods of prolonged heavy rain. This effect is clearly visible in Fig. 2.11, which compares the seasonal trends in pH of two rivers at Glendye (granite) and Glenbuchat (quartz–biotite–norite) over a twelve-month period in 1984/85. The autumn of 1984 was abnormally wet in north-east Scotland.

Fig. 2.10. Effect of equilibrating atmosphere carbon dioxide concentration on the pH of river water from Glendye, north-east Scotland.

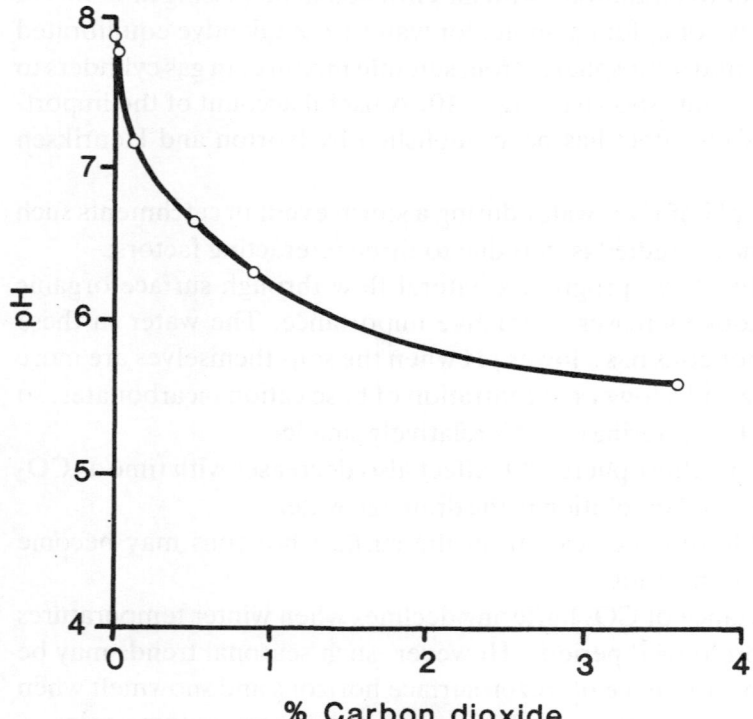

It might be expected that, if depletion of readily weatherable minerals is a prerequisite for substantial water acidification, streams which are susceptible to serious acidification during storms would generally be low in calcium. This is the case, but baseflow stream calcium concentration is not a particularly useful indicator of potential acidification risk because it takes no account of the changes in hydrological pathways during storms. Stormflow calcium concentration does, however. Thus a plot of the minimum calcium measured over a 12-month period provides a good indication of mean pH. Fig. 2.12 shows data plotted for nine streams studied by us and 21 streams studied by Pugh (pers. comm.). The encircled points are for the ten streams which had a minimum pH of <5.2 over the 12-month period.

It is worth pointing out that, for the data used to plot Fig. 2.12, the minimum calcium and minimum pH did not occur on the same date. When pH is plotted against calcium concentration for the mid-November sampling for the 30 streams (Fig. 2.13), on no occasion did the latter parameter fall below 0.6 $\mu$g ml$^{-1}$. This may reflect a high contribution from overland flow on this particular occasion. As a result of biological recirculation, calcium tends to be higher in soil very close to the surface than in underlying peat. Complex behaviour such as this makes it difficult to use water analysis from spot samples to predict susceptibility to acidification for rivers.

For lakes, calcium concentration provides a good indication of acidifi-

Fig. 2.11. Seasonal pH trends for rivers at: ———— Gendye (granite) and ---- Glenbuchat (quartz–biotite–norite) over a 12-month period in 1984/85.

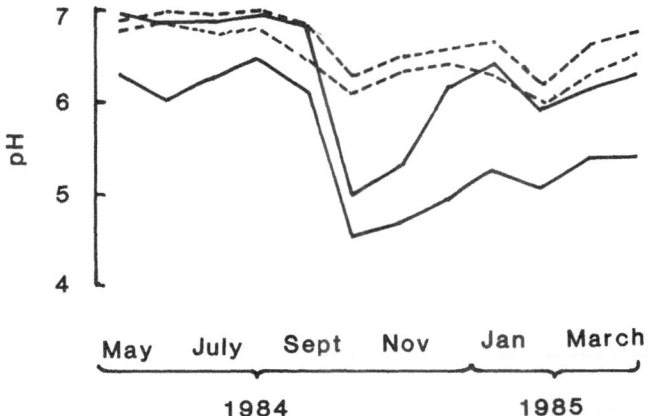

cation (Henriksen, 1979). This is because the quantity of water draining into the lake is greatest when pH and calcium are lowest. Thus lakes provide a natural weighting to favour stormflow water, and tend to reflect both depletion of readily weatherable minerals and hydrological pathway effects.

## Snow and snowmelt

Snow and snowmelt have already been mentioned at various points in this chapter. There is no doubt that rapid melt of accumulated snowpacks may lead to very acidic drainage water. It has sometimes been assumed that the acidity may be explained in terms of unmodified precipitation passing unchanged into drainage channels at melt. Since dry deposition between snow events will fall upon the surface of the snow cover, the lying snow becomes progressively more acidic prior to melt. Moreover, it might be anticipated that the most polluted snow will melt first as the air temperature rises, causing a pulse of very acid melt-water. The evidence for the fractionating of ionic impurities in this way has been summarised by Overrein *et al.* (1980). Hydrogen ions may be

Fig. 2.12. Relationship between river water mean pH and minimum calcium for 30 streams in north-east Scotland, based upon monthly samples over a 12-month period. Filled circles = Pugh's data; open circles = our data. Encircled points are for streams with a minimum pH below 5.2.

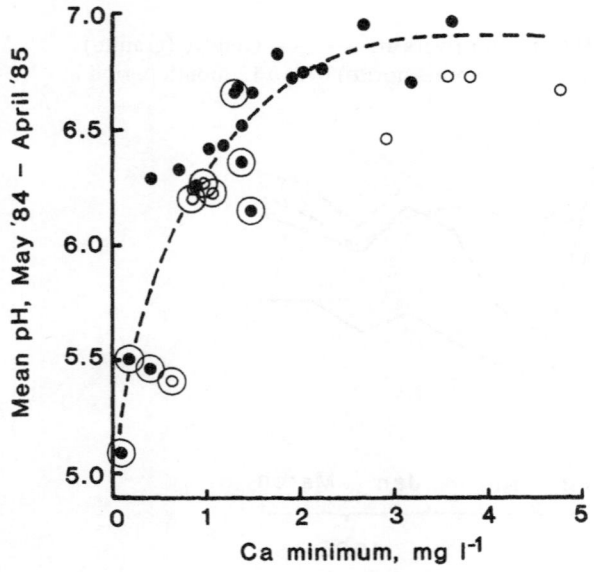

relatively concentrated ten-fold in early compared to late snowmelt. In our experience in north-east Scotland, meltwater tends to contact at least upper soil horizons, and the pH of the latter rather than that of the snow tends to govern drainage water pH, even when melt is rapid. If the melt is slower, water may penetrate to the mineral soil and may not produce especially acid streams. Even when air temperatures have been very low (e.g. *ca* −20 °C) for prolonged periods, soil horizons under an insulating snow cover may not be frozen, and water penetration to depth may occur. This happened in parts of Scotland after a severe winter in 1985/86. Thus particularly acid water is favoured by very rapid melt, by the same soil and topographic conditions which favour acidification during storms, and by frozen soil at a depth which confines water

Fig. 2.13. Plot of pH vs calcium concentration, for the 30 streams used to provide the data for Fig. 2.12, for the mid-November 1984 (abnormally wet) sampling.

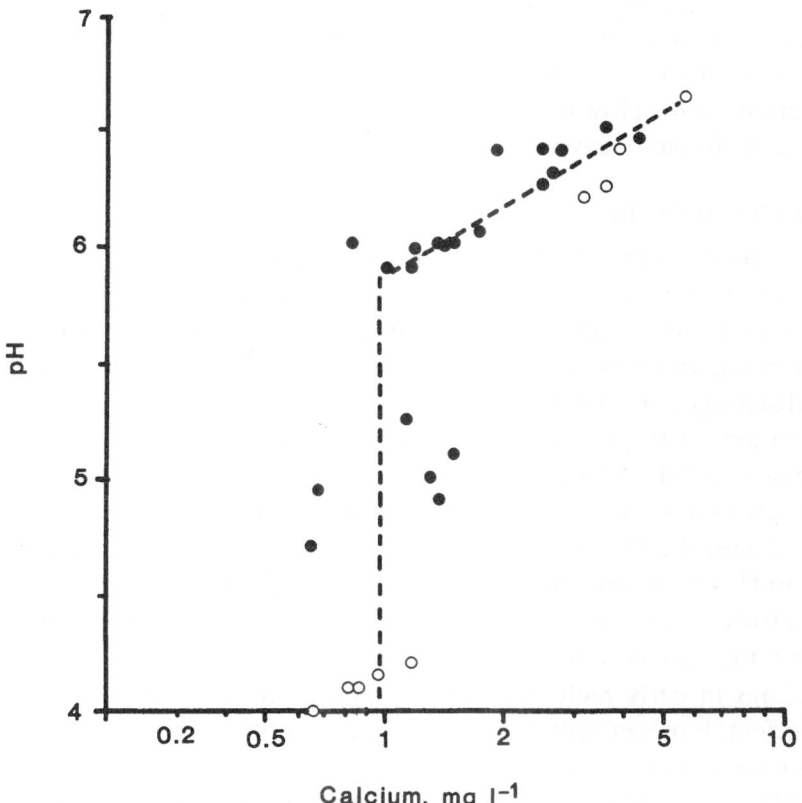

to more acidic surface horizons. Polluted precipitation is not a pre-requisite for severe stream acidification during snowmelt.

It would be incorrect to suppose that there is no mechanism for direct transfer of acidity from snow to lakes and, to a lesser extent, streams. Large shallow lakes may receive substantial direct inputs of snow (and of course rain). Moreover, recent work by Flower, Miller and Cresser (unpublished results) has shown that snow blowing downslope may be very significant. The amount of snow remobilised in this way depends upon the interaction between wind speed, snow packing, land use (veg-etation cover) and local topography. Empirically, blown snow contri-butions are difficult to measure precisely.

Where the outcropping bare rock in a catchment is substantial, the disadvantages of snowmelt are even greater. Interaction with rock sur-faces does modify precipitation chemistry to some extent, especially if the rock is lichen-covered (see e.g. Dahl *et al.*, 1979) and in the case of the first water washing over the rock surface. However, rapid melt of snowpack overlying rock leads to rapid runoff on to soil on lower slopes, with relatively little modification of the runoff solute chemistry *en route*. This extra water loading on lower slopes greatly increases the probabil-ity of generation of rapid throughflow through surface organic horizons or of overland flow. How much more stress the snowmelt imposes com-pared to a prolonged heavy storm depends upon the speed of the melt.

## Frozen soils

Marked changes in catchment hydrological pathways may occur during snowmelt if water flows over still-frozen ground, or laterally through surface soil horizons over still-frozen subsoils. In upland areas this may result in marked changes in water chemistry for a specified stream discharge. As mentioned in the previous section, early melt-water contains a disproportionate amount of the total solute load. Johannessen and Henriksen (1978) showed by both laboratory and field experiments that 50–80% of the pollutant load stored in the snowpack is released with the first 30% of the meltwater. If 'stored acid' reached streams and lakes, it could be very damaging at the hatching stage of fish species. However, we are of the opinion that the meltwater invariably reacts with the surface soil horizon. It is quite conceivable that the high acid loadings in early meltwater flushes in a polluted area could still have an effect, but primarily by increasing the acidity of the surface soil horizons over a period of years.

Edwards *et al.* (1986) have recently identified an additional factor which complicates detailed interpretation of changes in water chemistry during snowmelt. Leachates resulting from simulated rain on columns of upland soil previously stored at either 4 °C or − 12 °C for one week were considerably different. The leachates from the frozen soils showed substantial increases in concentrations of Ca, Mg, Na, K, Fe, Al, Si and total organic carbon. The changes were particularly dramatic and prolonged for Fe, Al and K. The effect was attributed to the rupture of plant cells and soil microfauna as intracellular fluid expands on freezing. This little-studied aspect is worthy of further investigation. So far attention has been largely confined to nitrate flushes which occur after periods of freezing of soil (Likens *et al.*, 1977). It has been suggested that freeze/thaw cycles promote nitrification and thus nitrate mobilisation (Arefyeva and Kilesnikof, 1964; McGarity and Rajaratnam, 1973).

## References

Amerman, C. R. (1965) The use of unit-source watershed data for runoff prediction. *Water Resources Research*, **1**, 499–507.

Anderson, M. G. and Burt, T. P. (1978) The role of topography in controlling throughflow generation. *Earth Surface Processes and Landforms*, **3**, 331–44.

Arefyeva, Z. N. and Kolesnikof, B. D. (1964) Chemistry and biochemistry dynamics of ammonia and nitrate nitrogen in forest soils of the Transurals at high and low temperatures. *Soviet Soil Science*, **3**, 246–60.

Atkinson, T. C. (1978) Techniques for measuring subsurface flow on hillslopes. In *Hillslope Hydrology*, Kirkby, M. J., ed., Wiley, New York, 73–120.

Ball, D. F., Mew, G. and MacPhee, W. S. G. (1969) Soils of Snowdon. *Field Studies*, **3**, 69–107.

Bloch, M. R., Kaplan, D., Kertes, V. and Schnerb, J. (1966) Ion separation in bursting air bubbles: an explanation for the irregular ion ratios in atmospheric precipitations. *Nature* (London), **209**, 802–3.

Bown, C. J. and Heslop, R. E. F. (1979) *The Soils of the Country Round Stranraer and Wigtown*, Memoirs of the Soil Survey of Great Britain, Scotland, Macaulay Institute for Soil Research, Aberdeen.

Cadle, R. D., Wartburg, A. F. and Grahek, F. E. (1971) Proportion of sulphate to sulphur dioxide in Kilauea volcano fume. *Geochimica et Cosmochimica Acta*, **35**, 503–7.

Childs, E. C. (1957) The physics of land drainage. In *Drainage of Agricultural Land*, Luthin, J. N., ed., *Agronomy*, **7**, American Society of Agronomy, Madison, 1–78.

Cohen, Y. and Gordon, L. I. (1979) Nitrous oxide production in the ocean. *Journal of Geophysical Research*, **84**, 347–53.

Cox, R. A. and Sandalls, F. J. (1974) The photo-oxidation of hydrogen sulphide and dimethyl sulphide in air. *Atmospheric Environment*, **8**, 1269–81.

Crampton, C. B. (1967) The evolution of soils on the hills of south Wales, and factors affecting their distribution, and their past, present and potential use. *Welsh Soils Discussion Group Report No. 8*, 52–69.

Crampton, C. B. (1982) Podzolization of soils under individual tree canopies in south-western British Colombia, Canada. *Geoderma*, **28**, 57–61.

Creasey, J. (1984) *The Geochemistry of a Small Upland Catchment in North-East Scotland*. Ph.D. Thesis, University of Aberdeen.

Cullis, C. F. and Hirschler, M. M. (1980) Atmospheric sulphur – natural and man-made sources. *Atmospheric Environment*, **14**, 1263–78.

Cuttle, S. P. (1983) Chemical properties of upland peats influencing the retention of phosphate and potassium ions. *The Journal of Soil Science*, **34**, 75–82.

Dahl, J. B., Qvenild, C., Tollan, O., Christophersen, N. and Seip, H. M. (1979) Methodology of studies on chemical processes in water runoff from rock and shallow soil cover using radioactive tracers. *Water, Air and Soil Pollution*, 11, 179–90.

Dawson, H. J., Ugolini, F. C., Hrutfiord, B. F. and Zachara, J. (1978) Role of soluble organics in the soil processes of a podzol, Central Cascades, Washington, *Soil Science*, **126**, 290–6.

De Kimpe, C. R. and Martel, Y. A. (1976) Effects of vegetation on the distribution of carbon, iron and aluminium in the B horizons of Northern Appalachian spodosols. *Soil Science Society of America Journal*, **40**, 77–80.

Delmas, R. (1979) Sulphate in polar snow and ice. *International Symposium on Sulphur Emissions and the Environment*, Society of Chemical Industry, London, 72–6.

Delmas, R. and Boutron, C. (1978) Sulphate in antarctic snow – spatio-temporal distribution. *Atmospheric Environment*, **12**, 723–8.

Duce, R. A. and Hoffman, E. J. (1976) Chemical fractionation at the air/sea interface. *Annual Review of Earth and Planetary Sciences*, **4**, 187–228.

Dunne, T. (1978) Field studies of hillslope processes. In *Hillslope Hydrology*, Kirkby, M. J., ed., Wiley, New York, 227–93.

Edwards, A. C. and Cresser, M. S. (1985) Design and laboratory evaluation of a simple fractionating precipitation collector. *Water, Air and Soil Pollution*, **26**, 275–80.

Edwards, A. C., Creasey, J. and Cresser, M. S. (1984) The conditions and frequency of sampling for elucidation of transport mechanisms and element budgets in upland drainage basins. *Hydrochemical Balances of Freshwater Systems*, IAHS Publication No. 150, Oxford, 187–202.

Edwards, A. C., Creasey, J., Skiba, U., Peirson-Smith, T. and Cresser, M. S. (1985) Long-term rates of acidification of UK upland acidic soils. *Soil Use and Management*, **1**, 61–5.

Edwards, A. C., Creasey, J. and Cresser, M. S. (1986) Soil freezing effects on upland stream solute chemistry. *Water Research*, **20**, 831–4.

Evans, L. J. (1980) Podzol development north of Lake Huron in relation to geology and vegetation. *Canadian Journal of Soil Science*, **60**, 527–39.

FitzPatrick, E. A. (1980) *Soils: Their Formation, Classification and Distribution*. Longman, London, pp. 288.

Fowler, D., Cape, J. N. and Leith, I. D. (1985) Acid inputs from the atmosphere in the United Kingdom. *Soil Use and Management*, **1**, 3–5.

Gerlach, T. M. and Nordlie, B. E. (1975) Carbon–oxygen–hydrogen–sulphur gaseous systems, parts I–III. *American Journal of Science*, **275**, 353–76, 377–94 and 395–410.

Gilman, K. and Newson, M. D. (1980) *Soil Pipes and Pipe Flow – A Hydrological Study in Upland Wales*, Geobooks, Norwich.

Goldich, S. S. (1938) A Study in rock weathering. *Journal of Geology*, **46**, 17–58.

Granat, L., Rodhe, H. and Hallberg, R. O. (1976) The global sulphur cycle. In *Nitro-*

gen, *Phosphorus and Sulphur – Global Cycles*, *SCOPE Report No. 7, Ecological Bulletin (Stockholm)*, **22**, 89–134.

Henderson, P. (1982) *Inorganic Geochemistry*, Pergamon, Oxford.

Henriksen, A. (1979) A simple approach for identifying and measuring acidification of freshwater. *Nature* (London), **278**, 542–5.

Hewlett, J. D. and Hibbert, A. R. (1967) Factors affecting the response of small water-sheds to precipitation in humid areas. In *Forest Hydrology*, Sopper, W. E. and Lull, H. W., eds, Pergamon Press, New York, 275–90.

Hillel, D. (1971) *Soil and Water*, Academic Press, New York.

Hitchcock, D. (1976) Atmospheric sulphate from biological sources. *Journal of the Air Pollution Control Association*, **26**, 210–15.

Jarvis, M. C. and Duncan, H. J. (1976) Profile distribution of organic carbon, iron, aluminium and manganese in soils under bracken and heather. *Plant and Soil*, **44**, 129–40.

Johannessen, M. and Henriksen, A. (1978) Chemistry of snow meltwater. Changes in concentration during melting. *Water Resources Research*, **14**, 615–19.

Jones, J. A. A. (1979) Extending the Hewlett model of stream runoff generation. *Area*, **11**, 110–14.

Junge, C. E. and Werby, R. T. (1958) The concentration of chloride, sodium, potassium, calcium and sulphate in rain water over the United States. *Journal of Meteorology*, **15**, 417–25.

Kinniburgh, D. G. and Edmunds, W. M. (1984) *The Susceptibility of UK Groundwaters to Acid Deposition – Report to the Department of the Environment*, British Geological Survey, Wallingford, pp. 211.

Kirkby, M. J., ed. (1978) *Hillslope Hydrology*, Wiley, New York.

Kirkby, M. J. and Chorley, R. J. (1967) Throughflow, overland flow and erosion. *Bulletin of the International Association of Scientific Hydrologists*, **12**, 5–21.

Langbein, W. B. and Iseri, K. T. (1960) General introduction and hydrologic definitions. *Manual of Hydrology, Part 1: General Surface-Water Techniques*, US Geological Water Supply Paper 1541-A, pp. 29.

Lee, R., Bache, B. W., Wilson, M. J. and Sharp, G. S. (1985) Aluminium release in relation to the determination of cation exchange capacity of some podzolized New Zealand soils. *The Journal of Soil Science*, **36**, 239–53.

Likens, G. E., Bormann, F. H., Pierce, R. S., Eaton, J. S. and Johnson, N. M. (1977) *Biogeochemistry of a Forested Ecosystem*. Springer-Verlag, New York.

Lindsay, W. L. (1979) *Chemical Equilibria in Soils*, Wiley-Interscience, New York.

Linsley, R. K. (1967) The relation between rainfall and runoff. *Journal of Hydrology*, **5**, 297–311.

Lovelock, J. E., Maggs, R. J. and Rasmussen, R. A. (1972) Atmospheric dimethyl sulphide and the natural sulphur cycle. *Nature* (London), **237**, 452–3.

Lowery, B., Kling, G. F. and Vomocil, J. A. (1982) Overland flow from sloping land: Effects of perched water tables and subsurface drains. *Soil Science Society of America Proceedings*, **46**, 93–9.

L'Vovich, M. I. (1979) *World Water Resources and their Future*, translated by Nace, R. L., American Geophysical Union, 415 pp. Cited in Shaw, E. M. (1983) *Hydrology in Practice*, Van Nostrand Reinhold, England.

McCaig, M. (1983) Contributions to storm quickflow in a small headwater catchment – the role of natural pipes and soil macropores. *Earth Surface Processes and Landforms*, **8**, 239–52.

McGarity, J. W. and Rajaratnam, J. A. (1973) Apparatus for the measurement of the losses of nitrogen as gas from the field and simulated field environments. *Soil Biology and Biochemistry*, **5**, 121–31.

Mitchell, B. D. and Jarvis, R. A. (1956) *The Soils of the Country Round Kilmarnock*, memoirs of the Soil Survey of Great Britain, Scotland, HMSO, Edinburgh.

Mosley, M. P. (1982) Subsurface flow velocities through selected forest soils, South Island, New Zealand. *Journal of Hydrology*, **55**, 65–92.

Nguyen, B. C., Gaudry, A., Bonsang, B. and Lambert, G. (1978) Reevaluation of the role of dimethyl sulphide in the sulphur budget. *Nature* (London), **275**, 637–9.

Norton, S. A. and Henriksen, A. (1983) The importance of $CO_2$ in evaluation of effects of acidic deposition. *Vatten*, **39**, 346–54.

Ottar, B. (1978) OECD study of long-range transport of air pollutants. *Atmospheric Environment*, **12**, 445–54.

Overrein, L. N., Seip, H. M. and Tollan, A. (1980) *Acid Precipitation – Effects on Forest and Fish*, Final report of the SNSF project, 1972–1980, Oslo.

Ovington, J. D. (1965) Nutrient cycling in woodlands. In *Experimental Pedology*, Hallsworth, E. G. and Crawford, D. V. eds, Butterworth, London, 208–15.

Reid, J. M. (1979) *Geochemical Balances in Glendye, an Upland Catchment in Grampian Region*, Ph.D. Thesis, University of Aberdeen.

Reid, J. M., MacLeod, D. A. and Cresser, M. S. (1981) Factors affecting the chemistry of precipitation and river water in an upland catchment. *Journal of Hydrology*, **50**, 129–45.

Rice, H., Hochumson, D. H. and Hidy, G. M. (1981) Contribution of anthropogenic and natural sources to atmospheric sulphur in parts of the United States. *Atmospheric Environment*, **15**, 1–9.

Roberts, E. (1958) *The County of Anglesey – Soils and Agriculture*, Memoirs of the Soil Survey of Great Britain, HMSO, London.

Rudeforth, C. C., Hartnup, R., Lea, J. W., Thompson, T. R. E. and Wright, P. S. (1984) *Soils and their Use in Wales*, Soil Survey of England and Wales Bulletin No. 11, Harpenden.

Rutherford, G. K. (1967) A preliminary study of the composition of precipitation in S.E. Ontario. *Canadian Journal of Earth Sciences*, **4**, 1151–60.

Skiba, U., Peirson-Smith, T. J. and Cresser, M. S. (1986) Effects of simulated precipitation acidified with sulphuric and/or nitric acid on the throughfall chemistry of Sitka spruce (*Picea stichensis*) and heather (*Calluna vulgaris*). *Environmental Pollution (B)*, **11**, 255–70.

Sugawara, K. (1967) Migration of elements through phases of the hydrosphere and atmosphere. *Chemistry of the Earth's Crust*, **2**, 501–10.

Tabatabai, M. A. (1985) Effect of acid rain on soils. *CRC Critical Reviews in Environmental Control* **15**, 65–110.

Thompson, T. R. E. and Loveland, P. J. (1985) The acidity of Welsh soils. *Soil Use and Management*, **1**, 21–4.

Thornton, J. D. and Eisenreich, S. J. (1982) Impact of land-use on the acid and trace element composition of precipitation in the north central U.S. *Atmospheric Environment*, **16**, 1945–55.

Ugolini, F. C., Minden, R., Dawson, H. and Zachara, J. (1977) An example of soil processes in the *Abies amabilis* zone of central Cascades, Washington. *Soil Science*, **124**, 291–302.

Vines, R. G. (1984) Rainfall patterns in the eastern United States. *Climatic Change*, **6**, 79–98.

Weyman, D. R. (1975) *Runoff Processes and Streamflow Modelling*. Oxford University Press, London.

Whipkey, R. Z. and Kirkby, M. J. (1978) Flow within the soil. In *Hillslope Hydrology*, Kirkby, M. J., ed., Wiley, New York, 121–41.

Williams, B. L., Cooper, J. M. and Pyatt, D. G. (1978) Effects of afforestation with *Pinus contorta* on nutrient content, acidity and exchangeable cations in peat. *Forestry*, **51**, 29–35.

Wilson, M. J. (1967) The clay mineralogy of some soils derived from a biotite-rich quartz–gabbro in the Strathdon area, Aberdeenshire. *Clay Minerals*, **7**, 91–100.

Wilson, M. J. (1975) Chemical weathering of some rock-forming minerals. *Soil Science*, **119**, 349–55.

# 3

○ ○ ○ ○ ○ ○ ○ ○ ○ ○ ○ ○ ○ ○ ○ ○ ○ ○ ○ ○

## Anthropogenic influences on acidification processes

Chapter 2 concentrated upon the natural ecological processes which regulate the extent of the acidification of freshwaters. Many of these processes are clearly susceptible to modification as a result of human activities, and these anthropogenic influences constitute the subject matter of this chapter. It should be realised from the outset that numerous factors other than the presence of acidic pollutants in the atmosphere and precipitation must be considered in this context. Other pollutants which may lead to soil and, in the long term, drainage water acidification, most notably ammonia (van Breemen *et al.*, 1982, 1983; van Breemen and Jordens, 1983), may be equally important in some instances. Changes in land use may influence drainage water pH in many ways, a point stressed quite emphatically by Rosenqvist (1977, 1978) early on in the acid rain debate, and raised increasingly in more recent overviews of soil acidification (e.g. Krug and Frink, 1983; de Vries and Breeuwsma, 1984; Rowell and Wild, 1985).

Land use changes may modify both the quantities of acidifying pollutants deposited and the fate of these pollutants, quite apart from any direct effect of cultivation techniques or different land use patterns upon drainage water acidity. It seems logical therefore to discuss the possible consequences of modifications to agricultural or forestry practices in relation to water acidification prior to considering the impact of acidifying pollutants, even although the latter has received appreciably more attention from researchers. Because so much effort has been directed towards the study of forest ecosystems, afforestation and forest clearance are considered first.

### Afforestation and forest clearance

The possible effects of tree growth upon the acidity of water draining from afforested watersheds has been discussed briefly in Chap-

ter 2. It is widely accepted that coniferous forest growth or forest regen-
eration on a clear-felled or burned forest site may lead, over a few years,
to the creation of acidic surface organic horizons. The evidence for this
has been succinctly summarised by Krug and Frink (1983), and need not
be considered again here. Suffice it to say that the production of surface
organic horizons of variable thickness with a pH appreciably below that
of the previous surface horizon is perfectly feasible, and underlying
mineral soil too may be acidified as a result of tree growth (see e.g.
Skeffington, 1983). Where the catchment hydrology is such that rapid
throughflow or return flow (see Fig. 2.3, p. 24) make a substantial contribution to stream discharge during storms, acidification of the stream
water to some extent must be unavoidable, even if the incident precipitation is not particularly acidic. Harriman and Morrison (1981) compared the chemistry of streams draining moorland and forested catchments in the Loch Ard area of central Scotland, and found the moorland
stream annual mean pH was 5.40, compared to the forest stream value
of 4.34. They suggested that these observations could be attributed to
the larger surface area, and hence greater collection efficiency of trees
for marine salts and gaseous pollutants. When discussing these findings,
Miller (1984a) commented that trees appeared somehow to 'facilitate
the passage of pollution-derived acidity into streams'. A probable
explanation is that the base saturation and pH of the soil surface organic
horizons is lowered by the forest growth and litter fall, and equilibrating
water draining laterally through or over these horizons is more acidic
regardless of the rainfall acidity. As discussed later, precipitation of
abnormally low pH may exacerbate this situation. Moreover, the
additional high input of neutral salts of maritime origin trapped by the
trees could lower the pH of water draining from such organic horizons.
This effect is considered later.

Miller (1985a) has described the association between afforestation
and stream-water acidification as a 'uniquely British feature' of the acid
rain debate. It is difficult to obtain really conclusive evidence for the
acidification of water as a result of forest growth. Areas allocated to
forestry frequently tend to be those with marginal upland soils, often in
regions with thinnish soils that are naturally acidic, with heavy precipitation and with relatively steep slopes. Such areas quite naturally tend
to have acidic drainage waters, particularly during heavy storm events
(see Chapter 2). For a valid direct evaluation of the afforestation effect,
it is necessary to find sites with near identical climate, soil parent
material and pedogenesis, hydrological properties, surface and bedrock
topography and prior land use history. Furthermore, an effect can only

be expected if the catchment hydrology is such that rapid throughflow, return flow, and possibly overland flow contribute significantly to total discharge. The alternative approach, finding an adequately uniform moorland site and planting trees on part of it is fundamentally very attractive, but would produce results only after a very long-term study. The reverse approach, clear-felling one of a pair of carefully matched afforested sites, also tends to yield results rather slowly. Such an approach is currently underway as part of a British Geological Survey project at Plynlimon in Wales. The effect of clear-felling was also included in the Hubbard Brook study (Likens *et al.*, 1977).

## Summary of effects of forest clearance/ afforestation

The influence of tree growth upon soil properties and hence upon water acidification was considered briefly in Chapter 2. It is appropriate to consider these effects again here, but this time in more detail, and with a view to assessing the impacts of large-scale planting or clearance.

*Changes in the hydrological cycle*

Tree growth or removal may influence the hydrological cycle in a number of diverse but interacting ways. As a general rule, in areas where there is a serious risk of freshwater acidification, soil pH tends to increase with depth down the soil profile. Thus the deeper water penetrates down the profile, the less acid the final drainage water is likely to be. The question which must be asked then is: How might forest growth change drainage water pathways to a lake or stream? Miller (1985a) commented that the most attractive hypothesis to explain the effect of afforestation on stream-water acidity was that changes in soil physical features, associated either with initial drainage improvements or with subsequent growth, were such that drainage water was less likely to enter soil horizons rich in relatively unweathered minerals, or residence time in such horizons was too short for complete neutralisation. Open channel flow in drains covered in very acidic litter deposits could undoubtedly have an acidifying effect on water during heavy storms. The reduced residence time hypothesis seems less plausible, since the ion exchange reactions regulating water pH are very rapid, the base cations displaced from exchange sites by $H^+$ being replaced by geochemical weathering between storms when residence time is longer. In the absence of a drainage network being established prior to planting, there appears to be little evidence to suggest that water drains more rapidly from a forested site than from a moorland site. Indeed, bearing in mind the greater interception and evapotranspiration losses from a forest canopy, especially for an evergreen coniferous species (see e.g. Jarvis *et al.*, 1983; Skeffington, 1983), rapid throughflow mechanisms might be expected to be less likely from the forest site on the basis that less water passes through the soil. On the other hand, in some areas, drying out of forest soils may be considerable in summer (Miller, 1984a) when appreciable water stress may occur (Jarvis *et al.*, 1983). This could lead to contraction of organic matter, particularly away from large lateral roots, and the development of very extensive interconnecting pipe-flow drainage systems through which water could drain at a rate approaching open channel flow (McCaig, 1983). Such water could become very acidic, and the adjacent organic soil could remain hydrophobic well into autumn. We have observed such behaviour in upland moorland catchments in north-east Scotland after an abnormally dry summer. In the absence of any conclusive evidence, such a hypothesis must remain speculative. Certainly an investigation of the macrostructure developing in forest compared to moorland soils in such upland areas in dry summers could provide useful information (FitzPatrick *et al.*, 1985).

*Deforestation and the possibility of placon formation*
As mentioned in Chapter 2, the presence of indurated horizons or placons (iron pans) may facilitate rapid downslope lateral flow through surface organic horizons (Tilsley, 1977) and hence contribute substantially to freshwater acidification during prolonged heavy rain. Placons are most frequently associated with maritime or moist alpine climatic conditions. They are generally only found where the soil parent material is freely drained, the temperature is low, and leaching is substantial. Active placon formation is still underway in much of Newfoundland and the western Scottish Highlands, although relic placosols are more widespread (Tilsley, 1977). In Scotland, placons which are substantial and continuous tend to be found under *Calluna* moorland. It is possible that *Calluna* moorland expansion over deforested areas could have led in the past to substantial placon formation and surface water acidification over a time scale of decades rather than centuries. Pan formation may also be encouraged by reduced grazing, as discussed later. Once formed, the presence of the placon often leads to the development of poorly drained peat horizons which are naturally very acidic. As far as we are aware, no

detailed study has yet been made of the probable link between the incidence and extent of iron pan formation and surface water acidification. Such a study could be very fruitful.

Re-establishment of forest on moorland sites which have well established placons may have a number of effects. If ploughing ruptures the pan, drainage to more basic soils may be improved, but on the other hand drying and oxidation of surface organic horizons could still cause very acidic pulses during storms.

*Changes in snowmelt and snowblow patterns*
Snowmelt is widely recognised to play an important role in freshwater acidification (see e.g. Goldstein *et al.*, 1984, and Chapter 2). Unfortunately our knowledge of hydrological behaviour during snowmelt is inadequate to allow precise comment upon the processes occurring and the likely effects of afforestation. Crucial factors are the rate of thawing and the relative amount of water during snowmelt which flows laterally downslope through or over (but contacting) surface organic horizons. Such water may be acidified to a considerable extent, regardless of the pH of the snow. Booty and Kramer (1984) suggested that this was one of the limitations to the applicability of their otherwise very successful model for acidification in a forested watershed in Ontario. The pH of lake water during snowmelt may be buffered by mixing of the acid melt-waters and the bulk of the water in the lake, unless mixing is impeded as a result of thermal stratification. The latter may be particularly pronounced under ice cover (Goldstein *et al.*, 1984). Streams draining lakes with frozen surfaces may therefore show serious acidification pulses during melt periods (Goldstein *et al.*, 1984).

Intuitively it seems highly probable that hydrological conditions prevailing during snowmelt may be different in a forested compared with a non-forested catchment, particularly where the tree species are not deciduous. The crucial issue is to what extent the presence of the trees changes the hydrological pathways followed by the meltwater to a stream or lake, particularly if more water passes over or through the surface horizons. Clearly there is a probability of such water being further acidified if the pH of the surface horizons is lowered as a result of atmospheric pollution inputs. It would be possible, but not particularly fruitful, to speculate at length about possible changes in pathways and the mechanisms which might be involved. Detailed field studies are really necessary first, particularly on litter and surface horizon pH values in relation to pollution levels, precipitation inputs, plant species and soil parent material and soil age.

What is certain is that effects of changes in drainage water solute composition with time during snowmelt (Johannessen and Henriksen, 1978) or as a result of soil freeze/thaw cycles (Edwards *et al.*, 1986) may be modified as a consequence of changing land use. Sorensen (1984) was critical of the fact that Christophersen *et al.* (1984) suggested at a meeting of the Royal Society in London on acid rain that acid surges in streams were connected with the fractionation effect in early snowmelt. Certainly the role of surface soils, particularly after freeze/thaw, cannot be ignored. Rosenqvist (1978) has stated that from his investigations there were no indications that runoff water obtained its low pH as a function of precipitation pH except where the runoff was from glaciers. It should be remembered that, even if water flows over or through snowpack on middle to upper slopes of a drainage basin, it may still contact acidic surface horizons on lower slopes before reaching a stream or lake. It is possible, however, to envisage conditions occurring on rare occasions under which impermeable sheets of ice coat the vegetation surface, and water flows over the surface of the ice relatively unchanged.

The effects of blown snow (see Chapter 2) do not seem to have been considered with respect to freshwater acidification. Work in the Grampian Region of north-east Scotland (Flower, Miller and Cresser, unpublished results) has shown that substantial amounts of snow may sometimes be blown towards the lower slopes of a catchment. Such snow blowing into large, shallow lakes may, when coupled to direct snow inputs and if the lake is not frozen, contribute significantly to the acidification of the lake water if the snow contains acidic pollutants. This topic is discussed later when the fate of atmospheric acidifying pollutants is discussed. Suffice it to say here that tree cover or its removal may substantially change the extent of snowblow and its importance.

*Possible effects upon mechanical erosion*
The presence of mature forest may have a considerable stabilising influence upon soils in sloping upland areas. Clear-felling at such sites may therefore result in substantial soil erosion, expanding significantly the area of bare and lichen-covered rock, so that the actual timber-line may be several hundred metres lower than expected if replanting or

regrowth occur (Rosenqvist, 1978). Erosion losses may facilitate fresh-water acidification in many ways. Loss of the soil water storage capacity on upper slopes means that more water will pass relatively less changed to the soil that remains on the upper slopes. When the water eventually reaches soil, the soil profile may be reduced in depth. Acidification of the residual mineral soil may be rapid, both through increased leaching as a result of the greater amount of water percolating through the soil and also because of the greater probability of sulphate saturation as a consequence of the redistribution of incoming sulphate flux. The greater water flux through the soil increases the likelihood of rapid throughflow occurring, as does the reduction in evapotranspiration. The effects are summarised in Fig. 3.1. The final outcome would be an increase in acidification of drainage water, particularly during heavy storm events, to an extent which depended upon the extent of mechan-

Fig. 3.1. Simplified schematic representation of the possible effects of erosion after clear-felling on hydrological pathways. Thicker arrows represent increased contributions; a = surface runoff; b = throughflow through surface, organic-rich horizons.

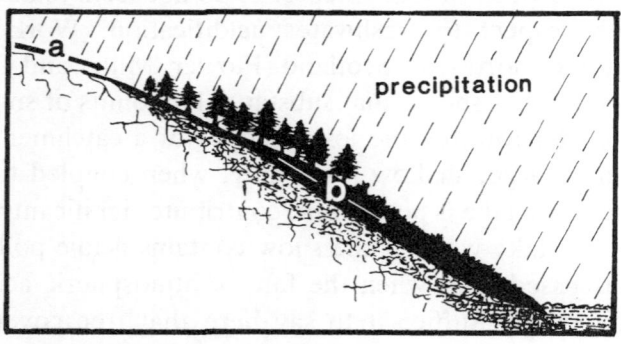

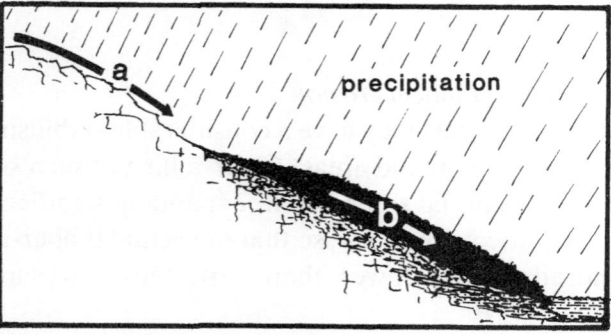

ical erosion and the change in the relative contributions to discharge of rapid throughflow and baseflow water. Such a phenomenon would obviously not be totally reversed by re-establishment of forest, which must in any case be on a reduced scale, within any useful time scale.

It is often suggested that clearance of indigenous hardwood forest by early man has resulted, over much of Western Europe, in the acceleration of natural soil acidification processes as a result of the establishment of heath vegetation (Dimbleby, 1962; Smith and Taylor, 1969; Hornung, 1985). Where the heath surface organic horizons were more acidic than the original forest surface horizons, this effect would further aggravate the problem described above. Vegetation species effects are considered again in the next section.

*Changes in soil properties following afforestation*
The impact of tree growth upon underlying soil profile chemical properties, particularly acidity, has been discussed in a concise but fairly comprehensive review by Hornung (1985). Clearly soil acidification may have corresponding effects upon drainage water acidity. It is appropriate therefore to consider here the possible impact of the tendency around the turn of the nineteenth century to replace indigenous hardwood or mixed forest with more rapidly growing conifers. A collation of the available data by Miles (1978) suggests that replacement of beech or oak by Norway spruce or Scots pine in many European countries causes a reduction in pH of up to one unit in surface horizons. Certainly the work of Matzner and Ulrich (1981) has demonstrated that the production of $H^+$ by Norway spruce was about double that by beech. Ovington (1953) concluded that, for the coniferous and hardwood tree species he studied in England, conifers tended to cause more surface horizon acidification than did hardwoods, but noted that initial soil conditions and management of a given stand could influence the extent of acidification, an observation borne out by other studies elsewhere; these have been summarised by Hornung (1985). As might be expected, acidification tended to be worse on soils that were already of low base status. In these soils geochemical weathering is least likely to be able to replace at a sufficient rate those base cations leached during storms or taken up by the tree.

From the above discussion it may be concluded that planting of conifers at upland sites might lead to drainage water acidification wherever lateral flow through surface organic horizons makes a significant contribution to total discharge and the mineral soil is of low base status, not uncommon characteristics for sites likely to be afforested. In areas susceptible to high inputs of acidic atmospheric pollutants, the high trapping efficiency of trees compared to moorland may further accelerate the acidification of surface soil horizons, and trapping of additional marine-derived aerosol may lower drainage water pH. These aspects are considered again later.

It should not be forgotten that *Calluna vulgaris* itself may produce very acidic surface horizons, when colonising degenerating grassland, for example (Grubb *et al.*, 1969). Indeed, Miles (1981) has reported that birch woodland establishment on some podzols which were previously under heather moorland may lead to increases in soil pH and base saturation and eventually to improvement of the soil to a cambisol.

*Soil changes in maturing forest*
In a review of the effect of acid rain upon soils, Tabatabai (1985) summarised evidence from the USA, Canada and Sweden which showed no obvious change in soil pH to depth over periods from 18 to 39 years. This suggests that a moderate degree of precipitation acidification is simply balanced by additional geochemical weathering in many soils (Miller, 1984a). However at one forested site in Sweden, Tamm and Hallbäcken (pers. comm.) found soil pH decreases of *ca* 0.5 units to a depth of 0.5 m over *ca* 40 years. We are of the opinion that $SO_4^{2-}$ anion adsorption may play a significant role in amelioration of the effect of the sulphuric acid component of acid deposition on soil acidification by limiting cation leaching. This is discussed in detail later in this chapter. Suffice it to say here that, if the presence of trees leads to greater $SO_4^{2-}$ trapping, $SO_4^{2-}$ saturation and profile acidification to depth become more likely. Anion adsorption differences may therefore be one aspect of the explanation for the apparent discrepancy between the pH stability of soils at different sites. Johnson and Henderson (1979) showed very convincingly that $SO_4^{2-}$ in a headwater in eastern Tennessee was regulated by anion adsorption at depth (Fig. 3.2), so that stream $SO_4^{2-}$ increased *ca* four-fold at high discharge, when water moving through surface organic horizons became important during storms.

Differences in the extent of geochemical weathering of soil minerals may also contribute to between-site variability. Because of the dominance of cation exchange reactions in the regulation of soil acidity, it should be expected that increasing the acidity of precipitation will merely cause a shift in cation populations on exchange sites in surface horizons until a new equilibrium is reached. Only where soil mineral weathering is very advanced should the soil pH be appreciably lowered at depth.

*Processes immediately following forest clearance*
It is appropriate to consider here the probable immediate effect of clear-felling upon drainage water acidity, apart from the possible erosion effects discussed earlier. If groundcover is negligible, evapotranspiration will fall dramatically and more water will pass through the soil. If lateral flow through organic horizons was already important, it will become increasingly so after deforestation, leading to further discharge acidification during storms. On the other hand, reduced trapping of dry deposition, both of acidic pollutants and neutral salts, may have a counteracting effect on drainage water acidity. Base cation recycling

Fig. 3.2. Typical $SO_4^{2-}$ hysteresis during a storm. Based upon the results of Johnson and Henderson (1979).

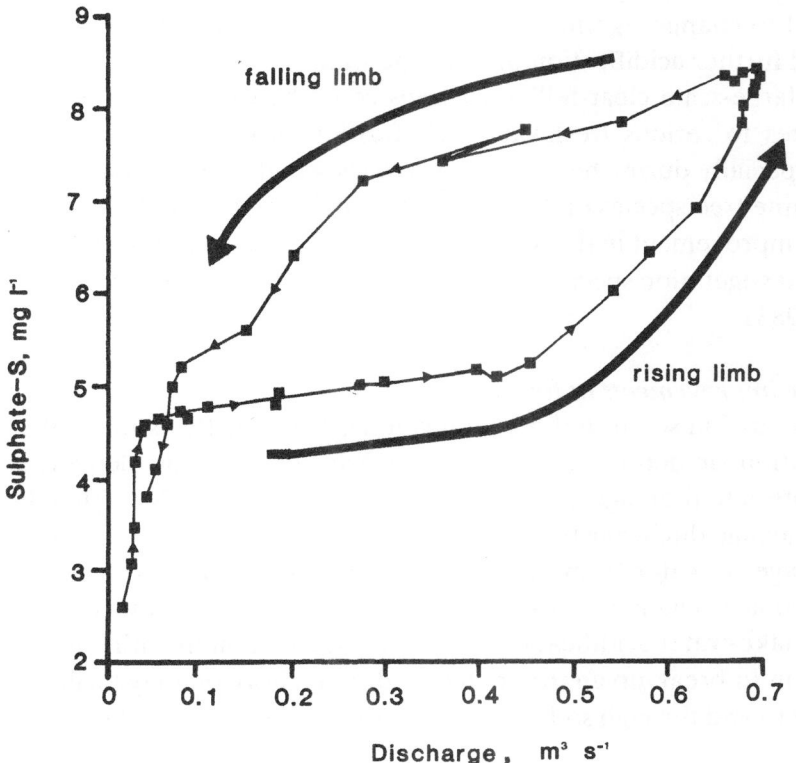

will temporarily be of little significance. At depth, however, base saturation in mineral soil might be expected to slowly increase slightly as a result of reduced base cation uptake. Unless the whole trees are removed, for coniferous species high levels of litter deposition will lead to further acidification or thickening of the litter horizon. Unless retained by the microbial biomass, nitrate and sulphate and associated cations will be mobilised through surface horizons whenever lateral flow prevails, leading to further lowering of the base saturation of surface horizons, and eventually to drainage water acidification. The sudden availability of large quantities of microbially degradable material, and possibly the increase in soil temperature, may lead to a surge in microbial activity and soil atmosphere carbon dioxide concentration. This will be countered to a large extent, however, by reduced root respiration. The net outcome on the $CO_2$-buffering effect (see Chapter 2) is difficult to predict. Mineralisation of organic N, S and P are all acid-producing processes (van Breemen *et al.*, 1983; van Breemen *et al.*, 1984). Removal of part or entire trees at harvesting obviously involves a major disruption of a long-term ecological cycle and a substantial soil acidifying effect (Hornung, 1985). Finally, snowmelt patterns may be expected to change significantly, almost certainly towards more rapid melt and further acidification of drainage water.

Thus large-scale clear-felling of trees could be expected to lead at some sites to serious freshwater acidification problems in the short term, especially during heavy storms and snowmelt. Re-establishment of the same tree species on the same site is unlikely to lead to any substantial improvement in this situation, nor is colonisation by acidophilic moorland vegetation species such as *Calluna vulgaris* (see e.g. Krug and Frink, 1983).

*Drainage improvements in forests*
In the discussion so far in this chapter, it has been tacitly assumed that afforestation or deforestation has been achieved with no deliberate modification to drainage and with negligible cultivation. Deep plough-ing or drainage ditch construction may, during storms, greatly facilitate the passage of water from acidic surface organic horizons directly to open drainage channels, resulting eventually in storm flow stream-water or lake-water acidification. In other instances, on the other hand, cultivation to break up an iron pan or indurated horizon may facilitate drainage to and through soil mineral horizons, thus reducing the extent

of water acidification which results from lateral flow through surface organic horizons.

When improved drainage leads to the elimination or substantial reduction in the volume of anaerobic soil horizons, the oxidation of both organic and inorganic soil components which may then proceed may lead to a sharp fall in pH of the drained soil through $H^+$ production, and hence to an increase in the acidity of the drainage water. Typical reactions are:

waterlogged soil      drained soil

$$4FeS + 9O_2 + 10H_2O \rightleftharpoons 4Fe(OH)_3 + 4SO_4^{2-} + 8H^+$$
$$4Fe^{2+} + O_2 + 10H_2O \rightleftharpoons 4Fe(OH)_3 + 8H^+$$
$$RNH_2 + 2O_2 \rightleftharpoons ROH + NO_3^- + H^+$$
$$RSH + H_2O + 2O_2 \rightleftharpoons ROH + SO_4^{2-} + 2H^+$$

It is worth pointing out that, even if no drains are built, afforestation also may lead to shifts in water-table and diminution of the volume of anaerobic zones on lower slopes, simply because of the greater evapo-transpiration from forests and increased interception of precipitation. This is thus another mechanism whereby the establishment of forest may lead to stream-water acidification in some areas. Whenever this mechanism is significant, variability of precipitation amount and temporal distribution must be taken into account when interpreting trends in drainage water acidity from year to year.

## Other land use changes

Aside from the effects of growth or clearance of forest discussed above, the other land use changes which should be considered include transition from arable or grassland to moorland and *vice versa*, and changes in animal types or stocking densities. Associated with increases or declines of agriculture are a range of cultivation practices. These changes, and their possible influence on freshwater pH, are discussed in this section.

*Animal grazing effects*

Rosenqvist (1978) has discussed the possible contribution of changes in agricultural practices in Norway to water acidification. Fig. 3.3, for example, shows the relative numbers of horses, cattle, sheep and goats in the Aust-Agder County in southern Norway in 1875 and 1976. Rosenqvist (1978, 1981) commented that decreased grazing was responsible for the spread of *Calluna* moorland and much increased forest growth on upland sites, many of which had very shallow soils following earlier deforestation and erosion. The increased moorland and forest cover at such sites could be instrumental in the acidification of rivers and lakes.

Conversely, van Breemen *et al.* (1982, 1983) and van Breemen and Jordens (1983) have discussed the impact of high levels of ammonia originating from animal manures upon soil acidification. They postulated that the ammonia reacts with atmospheric sulphur dioxide to yield ammonium sulphate, possibly on the surface of vegetation. Ammonium sulphate is washed into the soil where oxidation leads to the production of nitric and sulphuric acids. The overall reaction may be represented by the equation (van Breemen and Jordens, 1983):

$$(NH_4)_2SO_4 + 4O_2 \rightarrow 2HNO_3 + H_2SO_4 + 2H_2O$$

The oxidation apparently proceeded very rapidly even under very acidic soil conditions, leading to soil pH values as low as 2.8 in an extreme case. Van Breemen *et al.* (1982) calculated that, at two of their woodland sites, the ammonia loading was equivalent to approximately 800 mm of precipitation at pH 3. In this instance the potentially very high loading of acidifying pollutant was related to a very high animal stocking density.

It might be construed from the work of van Breemen and his colleagues that reduced stocking density could only have a favourable

Fig. 3.3. Changes in relative numbers of horses, cattle, sheep and goats in the Aust-Agder County of southern Norway between 1875 (hatched) and 1976 (solid). After Rosenqvist (1978).

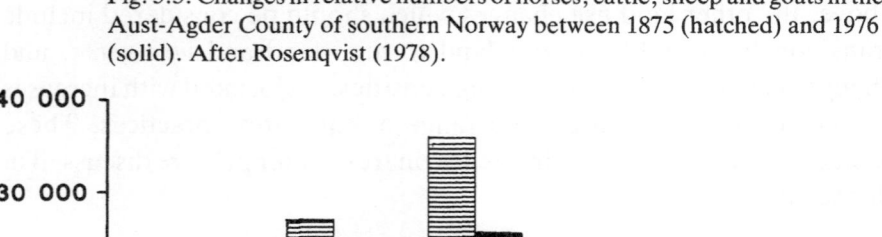

effect, by reducing the atmospheric ammonia loadings. However, the decreased pollutant loading effect could be largely offset by increased trapping in re-established forest. The hypothesis that land use changes could be a major factor contributing to freshwater acidification has been taken up and expanded by Krug and Frink (1983), who summarised the extensive evidence for organic surface horizons following afforestation.

*Evidence from diatom and pollen analysis*
Battarbee *et al.* (1985) have rejected the land use hypothesis for lakes in Galloway in south-west Scotland. They calculated the rate of acidification of Loch Enoch over the past 140 years from sediment core diatom analysis (Fig. 3.4), and compared the diatom data with *Calluna vulgaris* and Gramineae pollen analysis. The latter showed a distinct decline in the *Calluna* pollen contribution and an increase in Gramineae pollen.

Fig. 3.4. Rate of acidification of Loch Enoch in south-west Scotland deduced from diatom analysis. After Battarbee *et al.* (1985), reproduced with the author's permission, and by permission from *Nature*, **314**, 350 © 1985 Macmillan Journals Limited.

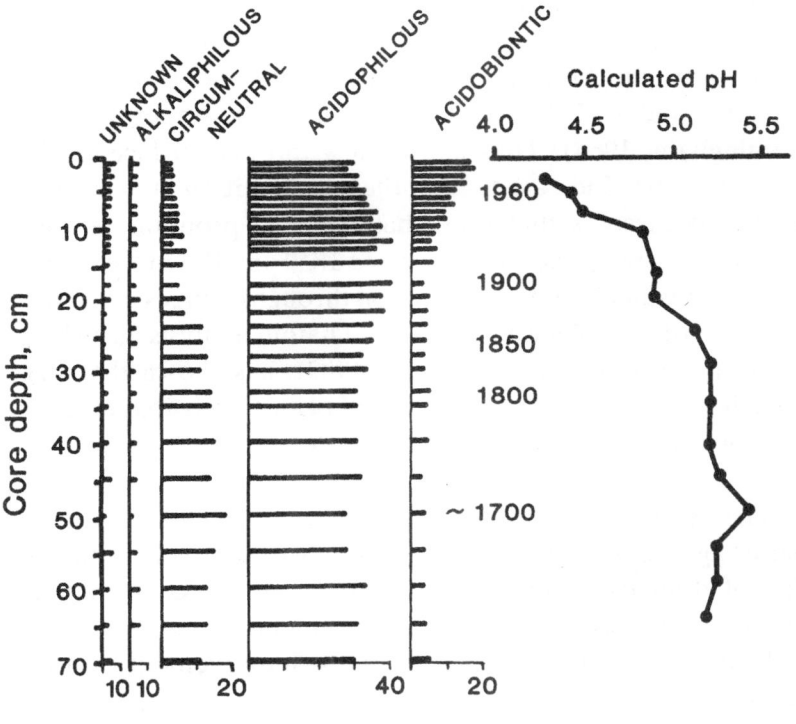

Battarbee *et al.* (1985) concluded that they could reject the reduced grazing land use hypothesis for their sites, since the *Calluna* was obviously being increasingly checked by grazing and repeated burning. Their conclusion warrants closer scrutiny, however. If it is accepted that the pollen analysis proves that the soil has *not* become progressively more acidic since *ca* 1700, it is then necessary to explain how water is passing through or over such a soil and still managing to become progressively more acidic. It is possible that *Calluna* is declining over the catchment as a whole, thus accounting for the pollen results, but not close to the edge of the lake in the zone which really matters as far as water acidification is concerned. Alternatively it cannot be ruled out that increasing grazing and heather burning has reduced the relative contribution of *Calluna* to total pollen, even although the soil is still becoming increasingly acidic. The circumstantial evidence presented by Battarbee *et al.* (1985) very strongly suggests that precipitation acidity is implicated in the lake-water acidification and certainly direct inputs of acid precipitation to the lake together with blown snow inputs may be important. We believe that soil acidification or modified hydrological pathways are also necessary to account for the rapid and extensive acidification of lakes such as Loch Enoch. Hydrological changes could occur as a result of the formation of an increasingly extensive and impermeable iron pan between 1800 and the present. Iron pans are known to form under *Calluna* moorland, and to be impermeable to water (Gimingham, 1964). Thus lateral flow through surface organic horizons would be increased. Another assumption inherent to Battarbee's hypothesis is that there has been no pronounced trend towards increased precipitation between *ca* 1850 and the present time, since this would result in water acidification through changes in the relative importance of hydrological pathways. Diatom analysis elsewhere, for example in Canada (Vaughan *et al.*, 1983), has shown that large fluctuations in water acidity may occur well prior to possible inputs of anthropogenic acidic pollutants on a large scale.

*Cultivation techniques and agricultural practices*
A number of agricultural practices may directly or indirectly influence the acidity of drainage waters. These include cultivation techniques such as shallow or deep ploughing, improvements to drainage, crop removal, heather burning on moorland sites, improvement of grassland, and the use of fertilisers and liming materials. The significance of any particular practice depends very much upon site characteristics and

climatic factors, and therefore it is worth considering each topic separately. It is also appropriate to consider here the possible impact of abandonment of each agricultural practice.

*Effect of ploughing*

On flat or gently sloping land in agricultural use, other than as permanent grassland, there is little evidence to suggest that drainage waters are ever likely to make a substantial contribution to the acidification of streams or lakes. This is primarily true because the fertility of such soils is generally maintained by liming to keep the soil pH above 6. Microbial activity in such soils is generally high, so that even in the temporary absence of a crop the soil atmosphere carbon dioxide concentration remains sufficiently large to ensure that the pH of drainage water is, after outgassing, 7 or higher. Ploughing such soils is of little consequence in the present context.

On sloping sites, however, ploughing may have a number of consequences. Deep ploughing which ruptures iron pans or indurated horizons which are relatively impermeable may substantially modify the hydrological pathways in a catchment and the extent of soil-pH-buffering mechanisms such as anion adsorption. This is so because drainage to greater depth through mineral soil horizons may become possible. Reduction in overland flow and rapid throughflow would then give rise to a reduction in the acidity peak observed during storm events. If, on the other hand, water penetration to depth is not facilitated after ploughing, then rapid flow from organic horizons to open channels may be increased, leading to increased stream-water acidification. Where catchments are very freely drained even prior to ploughing, as was the case for example for the coarse till at the Batchawana Lake Basin studied by Booty and Kramer (1984), ploughing is unlikely to have any significant effect aside from increased erosion. Regardless of whether or not drainage to depth is improved, downslope ploughing to produce drainage channels may lead to accelerated erosion, with a high risk of acidic soil components from upper slopes being deposited towards the bottom of a catchment. Drying out of previously saturated, anaerobic organic matter may, as suggested previously, lead to oxidation and acid production, and acid produced could pass rapidly to a stream or lake during heavy rain by open channel flow.

Lack of ploughing, for example through abandonment of agriculture, or a move towards direct-drilling techniques, may sometimes lead to substantial surface soil acidification. For example, Fig. 3.5 shows the influence of shallow cultivation and normal ploughing compared to a direct-drilled chalky boulder clay site at Boxworth in England after an eight-year trial (Chalmers, 1985). On shallow soils on slopes, or where there is any impediment to free drainage, this could lead to acidification of runoff during prolonged heavy rain. The converse, reduced acidification after ploughing of old grassland, is also possible on such sites.

*Drainage improvements*
The direct impact of improved drainage of upland sites, either via changes in hydrological pathways or increased oxidation of soil components, or both, has already been discussed in this chapter, and need not be discussed again here. Many minor secondary effects may also occur, however. For example, improved growth may lead to increased evapotranspiration, decreased water flow through the soil and increased litter production. Litter produced tends to accumulate in drainage trenches, and may contribute significantly to water acidification. Improved drainage of flatter, agricultural sites may also lead to increased oxidation of organic nitrogen and sulphur and inorganic species and soil acidification, but if the soil is limed regularly to maintain fertility this is unlikely to be a serious problem.

Fig. 3.5. Influence of cultivation technique on the pH profile of a chalky boulder clay soil. Filled circles = direct-drilled; open circles = shallow cultivation; open triangles = ploughed. After Chalmers (1985).

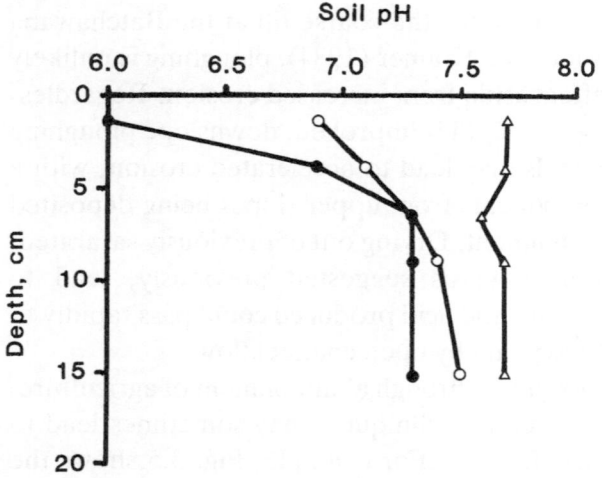

*Crop removal*

In the absence of harvesting, base cations taken up by plants are returned to the soil surface in plant litter. Crop removal however leads to removal of substantial amounts of base cations and thus contributes to permanent soil acidification (van Breemen *et al.*, 1983). For fast-growing conifers, mean annual soil acidification is highest when the trees are still relatively young, i.e. <20 years (Nilsson *et al.*, 1982). Regular harvesting of relatively young trees would therefore have the greatest soil acidifying effect. A crucial factor is the extent to which geochemical weathering (or fertiliser use) can replace base cations taken off site at crop removal. Clearly losses are greater when the whole tree is harvested, rather than just the trunk (Nilsson *et al.*, 1982). Soil acidification from cropping may be an order of magnitude or more greater than that attributable to acid deposition.

Gasser (1985) has recently reviewed the processes contributing to calcium loss from agricultural soils. He cited calcium outputs from harvesting of between 11 kg ha$^{-1}$ (potatoes) and 100 kg ha$^{-1}$ (red clover cut for hay). To replace the latter, limestone would have to be applied at 280 kg ha$^{-1}$. It should be remembered however that, because of poor growth, harvest-related decreases in soil pH would generally be less important on soils susceptible to pH fall.

From the above discussion it may be concluded that crop-removal-induced soil acidification is only likely to be important in the context of freshwater acidification when soil weathering is at an advanced stage and geochemical weathering is incapable of replacing lost base cations. The presence of an iron pan or impermeable indurated horizon would eventually increase the susceptibility of a site to freshwater acidification by limiting the mineral soil depth contributing to the maintenance of surface soil base saturation. As Tilsley (1977) has pointed out, the soils above and below a placon may be regarded as chemically isolated entities.

*Heather burning and forest fires*

Regular burning of moorland vegetation for management purposes has been a common practice in Britain, Sweden, Germany and Denmark in recent centuries (Gimingham, 1981), and is still commonplace in upland Britain. The purpose is primarily to encourage young growth on heath-land sites which are to be used for hunting or grazing. The effects in relation to water acidification may be far reaching in the short term and medium term (Starr, 1985). In normal to severe fires, around 65% of the carbon, 55% of the sulphur and 70% of the nitrogen in *Calluna* may be lost in smoke (Gimingham, 1972). The residual ash may have a significant liming effect upon the soil, especially if the fire is a severe one. Under very dry conditions, the surface of peaty horizons may itself burn, leading to loss of organic acids from the system. Coupled with the liming effect, this may cause a significant rise in surface horizon pH and thus tends to reduce water acidification. However, such an effect is only really likely to be substantial after more ecologically significant forest fires.

Some valuable data which shows the potential effect of severe forest fires has been presented by Rosenqvist (1981). Rosenqvist found that when an area of 4 km$^2$ of pine, spruce and heather forest at a site in Lifjell, Norway, was burned so severely that the humus horizons were lost, and the area was subsequently colonised by birch, willow herb and raspberry bushes, streams draining the burnt area had pH values of 5.8, 7.2 and 6.8, whereas those from the unburnt area gave river-water pH values of 4.5, 4.3 and 3.9. The soil at the site had evolved from quartzite, so presumably the improved site would eventually deteriorate back to give more-acidic organic surface horizons and drainage waters, on a time scale depending upon the underlying mineral horizon properties. Rosenqvist (1981) also cited comparable data for other sites in Norway. The important point that he wanted to emphasise was the vital role played by surface acid organic horizons at many of the sites at which river-water or lake-water acidification is observed. The study is no less valuable in the interpretation of the possible effects of burning as a management practice however.

Several other factors must be taken into account when considering the outcome of heather burning, especially if the burn is not particularly hot and the soil humus horizons have been scarcely affected. Interception by vegetation is reduced and evapotranspiration also temporarily falls, leading to more water draining through or over the soil. Uptake of nitrate by vegetation may also be much reduced, facilitating base cation leaching. Structure in the soil may be modified, causing changes in hydrological pathways. If a high temperature burn extends over the top of a hill, and no live roots are left, recolonisation may be slow, leading to substantial erosion of the soil on upper slopes. Sometimes we have noticed organic surface horizons remaining somewhat hydrophobic for months or even years, leading to large areas of increased surface runoff. Lack of vegetation cover increases the risk of snow being blown downhill.

The complex interacting effects described above mean that prediction of the precise effects of fire is only feasible for really hot fires in which humus horizons are destroyed. For more typical and moderate moorland burns, some of the secondary effects tend to increase the risk of acidification, offsetting to some extent the potential benefits of the liming effect of *Calluna* ash.

*Use of fertilisers and liming materials*

One aspect of the debate on the causes of acidification of freshwaters on which there is general agreement is that the use of nitrogen fertilisers may contribute significantly to soil acidification at some locations. Non-fertiliser inputs of ammonium-N in precipitation are generally only around 2–8 kg ha$^{-1}$ in Europe. For example, fluxes of 4.9–7.5 kg ha$^{-1}$ a$^{-1}$ have been reported for Czechoslovakia (Paces, 1985) and of 2.5–6.8 kg ha$^{-1}$ a$^{-1}$ for two catchments in the Grampian Region of north-east Scotland (Edwards *et al.*, 1985a). In areas with very high animal populations, such as the site in the Netherlands studied by van Breemen *et al.* (1982), the ammonium-N input in precipitation was 14.6 kg ha$^{1}$ a$^{-1}$. This value was swelled in throughfall to 42.4 kg ha$^{-1}$ a$^{-1}$ as a result of dissolution of dry deposited ammonium sulphate. Each mole of ammonium nitrified leads to the production of two moles of H$^{+}$ (van Breemen and Jordens, 1983) i.e.

$$NH_4^+ + 2O_2 \rightarrow NO_3^- + 2H^+ + H_2O$$
$$(NH_4)_2SO_4 + 4O_2 \rightarrow 2HNO_3 + H_2SO_4 + 2H_2O$$

The significance of this process has already been discussed in relation to natural atmospheric ammonia concentrations and animal stocking density changes. Clearly, fertiliser-N may have an even more substantial effect upon soil acidification. Over 11 million tonnes of N fertiliser was consumed in the USA in 1982, for example (Tabatabai, 1985). Tabatabai cited mean application rates for N of 44, 57 and 139 kg ha$^{-1}$ for wheat, cotton and corn, respectively, over a crop acreage of 1.38 $\times$ 10$^8$ ha. The chemical form of the fertiliser used is important, the overall reactions for urea and anhydrous ammonia being:

$$(NH_2)_2CO + 4O_2 \rightarrow 2NO_3^- + 2H^+ + H_2CO_3$$
$$NH_3 + 2O_2 \rightarrow NO_3^- + H^+ + H_2O$$

To put this in perspective, application of 100 kg of ammonium-N ha$^{-1}$ may add up to about one and a half times as much H$^{+}$ to soil as 1000 mm of pH 3 rain, or 14 times as much H$^{+}$ as 1000 mm of pH 4 rain. Some of this H$^{+}$ is of course offset by crop uptake and some by denitrification.

Nevertheless, one is led to the apparently irrefutable conclusion that typical fertiliser ammonia inputs could easily become of far more consequence to soil and water acidification than inputs of nitric and sulphuric acids in wet and dry deposition.

In a recent review of the processes causing losses of calcium from agricultural soils, Gasser (1985) pointed out that 1 kg urea-N, ammonia and ammonium sulphate required application of 3.6, 3.6 and 7.2 kg of lime ($CaCO_3$), respectively, to maintain soil base status. These figures are of course an approximation, as they do not take into account crop growth effects, but they serve as a useful starting point for discussion. Edwards *et al.* (1985b) reported geochemical-weathering-derived calcium losses of 17.2 and 20.2 kg $ha^{-1}$ $a^{-1}$ from two upland moorland catchments in north-east Scotland. Clearly application of substantial levels of ammonium-N, whether pollution or fertiliser-derived, must, if not accompanied by liming, either lead to a substantial increase in geochemical weathering rate or to rapid soil acidification. Paces (1985) reported a four-fold increase in total weathering of gneiss in catchments in Czechoslovakia as a result of agricultural practices.

For land in agricultural use which is cultivated regularly, the acidification of soils from fertiliser-N applications is likely to be of little consequence to water pH of streams and lakes in the short term, provided the soil is routinely limed to maintain its pH and base saturation. The greatest risk probably comes from fertilisation in forest and permanent grassland sites, where there is a natural tendency towards surface horizon acidification as discussed earlier and often no lime is applied (Department of the Environment, 1976). The extent of the problem which results depends upon the hydrology of the site, and the prevailing stage of growth and soil microbial activity. With N being applied to grassland in the UK at levels up to 400 kg $ha^{-1}$ and at an average rate of 190 kg $ha^{-1}$ in 1983 (Gasser, 1985), this factor must not be overlooked. Van Breemen and Jordens (1983) found that nitrification was rapid even in very acid woodland soils (pH 2.8–3.5). It should not therefore be assumed that declining bacterial activity and increasing fungal dominance of the microbial biomass will limit the nitrification effect in acid soils.

Discussion of the potential consequences of increased use of N fertiliser on soil and freshwater acidification are generally confined to the effects of $H^+$ production associated with nitrification. However, several other factors should also be taken into account. Substantial increases in crop yields lead to higher uptake of base cations, which may be removed with the crop. In forests, on the other hand, litter production may be significantly increased. Evapotranspiration losses may increase as a result of improved crop growth, and so on. Once again a complex series of interacting processes needs to be taken into account in a full analysis of the fertiliser effect.

Addition of any fertiliser with a relatively mobile cation/anion pair may also substantially lower drainage water pH in the short term through a cation exchange effect. A high concentration of ammonium or potassium, etc., in soil solution displaces $H^+$ and other cations from cation exchange sites on the soil matrix, immediately lowering the soil solution pH. Fig. 3.6, for example, shows how increasing the concentration of calcium, as the chloride, in the equilibrating solution lowers the pH down the podzol profile (Cresser *et al.*, 1986). Potassium from sulphate of potash or muriate of potash would show similar trends. Thus, unless a readily soluble fertiliser contains a neutralising anion, its use may significantly lower the pH of the drainage water. In particular

Fig. 3.6. Effect of calcium concentration (as the chloride) in equilibrating solution upon soil pH of samples from a podzol profile at Glendye, Grampian Region. Open circles = 0; filled circles = 5; open squares = 19; filled squares = 50 and open triangles = 100 μg ml$^{-1}$ calcium.

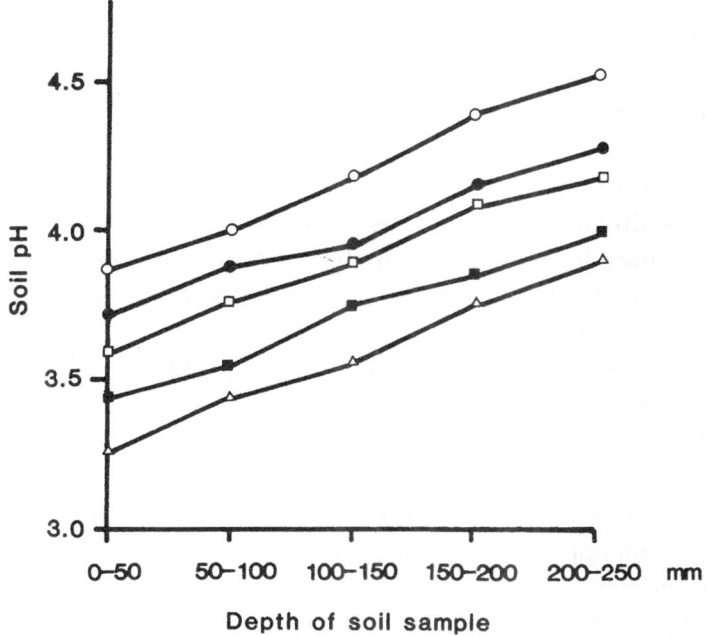

the pH of stream or lake water may be significantly lowered on upland sites during storms, but an effect may also occur at flatter sites if the fertiliser anion remains mobile.

If the fertiliser anion has a neutralising capability, however, the $H^+$ ions displaced from the exchange sites simply react with the counter anion; e.g.

$$\text{Exch}\underset{\diagdown H^+}{\overset{\diagup H^+}{}} + Ca^{2+} \rightleftharpoons \text{Exch–Ca}^{2+} + 2H^+$$

$$2H^+ + 2OH^- \rightleftharpoons 2H_2O$$

In this instance the acidifying effect of the fertiliser cations is ameliorated.

Liming of soils in catchments susceptible to freshwater acidification would clearly be beneficial, but not necessarily economically viable on the scale required as a curative measure. It may be easily shown that 1000 mm of rain at pH 4 would require 500 moles of lime $ha^{-1}$ to neutralise its mineral acid content, i.e. only *ca* 50 kg of calcium carbonate $ha^{-1}$. However, liming a band of soil to a pH of *ca* 6 might lead to a leaching loss of *ca* 1000 kg calcium carbonate $ha^{-1}$ $a^{-1}$ (Chalmers, 1985). Liming a strip of soil to a pH of *ca* 5 to 5.5 would lower the annual maintenance lime requirement four-fold, and might prove sufficient to alleviate some of the worst effects of acidification. Maintenance would be difficult in areas with very high rainfall such as the uplands on the west coast of Britain, where overland flow and return flow are much more prevalent.

### The importance of lower slopes

In upland areas, as already mentioned, overland flow, return flow and rapid throughflow may become increasingly important during heavy storms and snowmelt. It follows automatically from this that physico-chemical properties of soils on the lower slopes of drainage basins play a vital role in regulation of drainage water pH. Indeed, Miller (1984a) has even suggested that control of water passage through this zone may be an effective measure in reducing freshwater acidification. Tradition-ally there is a tendency to expect higher base saturation in the soils that constitute the lower slopes, because of the input of weathering-derived base cation fluxes from upper slopes. Such soils help buffer streams and lakes against the effects of the more rapid acidification which takes place on upper slopes. In areas where a high percentage of the precipitation falls as snow, the tendency towards downhill drifting may offset to some extent the benefits of inputs of cations from upper slopes.

The remains of abandoned crofts in Scotland suggest that for several centuries there was a tendency to favour agricultural management of lower slopes, where shelter is better, water easily available, and slopes are often more gentle. Furthermore, the soil may be more readily cultivated and may be more fertile. Upper, more exposed slopes would be confined to grazing. Any input of lime where a suitable liming material was available would be confined therefore to lower slopes. Even a relatively narrow band of such cultivated land could limit the extent of water acidification. Aside from the beneficial effects of any liming material, cultivation would retard the formation of well-defined, acidic, organic horizons, and might improve drainage to lower, mineral soil. Any stone and turf walls built to retain animals could minimise downslope snow drift and overland flow.

Abandonment of such settlements during the Scottish Highland clearances in the latter part of the eighteenth century as more and more land was used for sheep (Darling and Boyd, 1964; King and Nicholson, 1964) could have led to rapid soil surface horizon acidification, and hence contribute significantly to water acidification. The continued forest clearance to provide charcoal for an ever-growing number of iron-smelters (Ritchie, 1920) would have undoubtedly aggravated the situation still further, increasing stream discharge and, in the longer term, modifying catchment hydrology by iron pan formation under the *Calluna* moorland thus established. Even in an area such as the southern Scottish Highlands, with a very long history of intensive hill sheep farming (McVean, 1964) where high grazing has reduced *Calluna* cover, an iron pan will remain and stormflow water will continue to be acidic in the long term.

The above discussion concentrates upon contributing factors in Scotland, but many of the principles discussed could be applicable elsewhere in areas of comparable soil and climate.

*Modifications to natural water supplies*
So far in this chapter, only the indirect effects on water acidity of land use changes or changing agricultural practices have been considered. It should not be forgotten, however, that over recent centuries, man has often modified natural drainage patterns, for example by construction of dams or by pumping of stream or lake water for domestic water supplies. Dam construction raises the water-table and may increase the

amount of water draining into a lake through or over acid organic horizons. Conversely, pumping water out of a lake may lower the water-table and, in some instances, raise the water pH. The specific consequences of such activities must be considered in each particular case.

## Acidifying pollutants in the atmosphere

The reasons for the implication of atmospheric acidifying pollutants as a cause of freshwater acidification in the minds of the general public and by the media have been discussed in Chapter 1. The recent political and, in terms of funding, research ramifications of this association of ideas have been succinctly summarised by Sasanow (1985), and need not be considered here. The scientific community remains divided, however, about precisely how important these pollutants are. Authors such as Rosenqvist (1978, 1981) and Krug and Frink (1983), who have argued that natural acidification processes and land use changes predominate in many situations, have sometimes been strongly criticised, for example by Johnson *et al.* (1984). The counter argument stresses the importance of the nature of the inorganic acid anions associated with protons in water acidification.

We are of the opinion that the truth lies between the two extremes. Natural acidification can and does lead to very acid soils and correspondingly acid water under unfavourable hydrological conditions. The strong acid anion required often may be predominantly chloride, entering the system as sodium chloride of maritime origin. Theoretically $3.5~\mu g~ml^{-1}$ chloride in rain could lead to a drainage water pH of *ca* 4 if concentration (resulting from water loss by evapotranspiration) and dry deposition inputs are considered. On the other hand, entry to rivers and lakes of acid from precipitation with little neutralisation by soil is feasible under some circumstances. Furthermore, conditions definitely do sometimes occur under which acidifying pollutants lead to a reduction in soil pH and thus indirectly to a decrease in drainage water pH. The purpose of the following sections is to critically assess how acid deposition may influence drainage water acidity.

*Origins and distribution of atmospheric pollutants*

Franks (1983) summarised the major sources of $SO_2$ and $NO_2$ emission in the mid-1960s on a global scale. Of the 148 million tonnes of $SO_2$ emitted annually as a result of human activity, 104 million arose from coal combustion, 28.9 million from use of petroleum products, and 15.9 million from smelting and other sources. In the case of $NO_2$, coal burning only yields 27.3 million tonnes, whereas petroleum product use is responsible for 22.6 million tonnes. Examination of anthropogenic and natural sources of sulphur in the atmosphere suggests that, on a global scale, emissions from the two groups of sources are roughly comparable (Granat et al., 1976). For nitrogen, such a comparison is less reliable. Sulphur dioxide, hydrogen sulphide and various oxides of nitrogen are oxidised and hydrolysed to sulphuric and nitric acids in the atmosphere, at rates which depend upon the prevailing environmental conditions (Likens et al., 1972). Empirical observations and model calculations both suggest that nitric acid is formed more rapidly than sulphuric acid with distance from the emission sources, and that long-range transport of the former is therefore rather less (Rodhe et al., 1981).

Where the acidity of precipitation is thought to be a potential problem, it is largely sulphuric and nitric acids that are assumed to be responsible for lowering the pH below 5.65, the value for distilled water equilibrated with the carbon dioxide in the atmosphere (see e.g. Tabatabai, 1985). In practice, because of the natural sources of acidifying materials present in the atmosphere, 'unpolluted rain' would probably have a pH value of 4.9 to 5.0. As a result of fossil fuel exploitation, values in the pH range 3.5 to 4.5 are not uncommon in parts of the USA (Tabatabai, 1985) and much of Europe. Sometimes appreciably lower values have been recorded (Barrett et al., 1982).

In the UK, inputs of $H^+$ in precipitation only exceed 1 kg ha$^{-1}$ a$^{-1}$ in the very high rainfall areas of the north and west, such as the west central highlands of Scotland and parts of Cumbria (Fowler et al., 1985). For Britain, estimates of dry deposition of $SO_2$ suggest that it may contribute more to the total $H^+$ flux deposited than input in precipitation, with levels exceeding 2.4 kg ha$^{-1}$ a$^{-1}$ in the industrial areas of the Midlands and northern England and the major urban areas of the Scottish central lowlands (Fowler et al., 1985). The relative contributions of wet deposited and dry deposited sulphur in Europe tend to change with distance from the coast, wet deposition becoming relatively less important with distance southwards from the North Sea (Howells and Kallend, 1984). As in Britain, higher levels of precipitation with increasing altitude tend to increase the significance of wet deposition. Clearly the relative importance of dry deposition also decreases at a site more remote from pollutant sources (Granat, 1983).

Reliable long-term historical records of precipitation acidity are unfortunately rarely available. Maps have been published showing the large increase in acidity of rainfall over a large part of central Europe between 1956 and 1966 (Likens *et al.*, 1972). Brimblecombe and Stedman (1982) have pieced together information from the records of agricultural research stations in eastern North America, England and Belgium from the late-nineteenth century to demonstrate five to ten-fold increase in precipitation nitrate levels between then and now. Fortunately nitrate in rain was of interest to early workers because of its importance to crop growth. Paces (1982, 1985) has shown the substantial increases in concentrations of nitrate, sulphur, chloride and a range of base cations in water from the River Elbe in Czechoslovakia based upon analysis in November 1892 and November 1976 at similar discharge and in years which were climatically similar. Part of this increase must of course be attributed to anthropogenic influences other than atmospheric acidifying pollutants.

There is occasionally evidence produced that the acidity of precipitation at some localities is no longer rising, and may even be falling. For example, it has recently been shown that the pH of rain at Rothamsted and Saxmundham in England rose from 4.4–4.6 in 1969–73 to 4.8–4.9 in 1979–83, possibly in direct response to the energy price crisis. Sulphate flux at the same sites was apparently increasing, a fact attributed to increased rainfall (Goulding and Poulton, 1985). Edwards *et al.* (1985a) have shown that high seasonal precipitation leads to lower nitrate concentration, and it seems probable that a similar dilution of washout might be found for pollution-derived sulphate. Sulphate concentration changes may reflect increased ocean storminess in wetter years in a country with a maritime climate.

Evidence such as that discussed above, and records of annual amounts of fossil fuel combustion, quite clearly show substantial increases in precipitation contents of nitric and sulphuric acids from the early to the mid-twentieth century, with some evidence to suggest a more recent trend towards a decline in the levels of $SO_2$ (Watt Committee on Energy, 1984). That sulphuric and nitric acids constitute a major part of precipitation acidity is borne out by an analysis of precipitation chemical composition over a year or more by the statistical technique known as factor analysis (see e.g. Reid *et al.*, 1981; Edwards *et al.*, 1984a). The relationship between sulphate, nitrate and reciprocal of pH was clearly shown for catchments in the Grampian Region of Scotland.

Because of the seasonal nature of energy requirements in countries with a cool climate, there is often a marked seasonal trend in atmospheric concentrations of $SO_2$ and $NO_x$ and in sulphate and nitrate concentrations (Barrett *et al.*, 1982). The trend has been observed at several sites in Britain and in Scandinavia. The sulphate and nitrate precipitation maxima lag three months behind the maximum atmospheric $SO_2$ and $NO_x$ concentrations (December–January in the UK) (Barrett *et al.*, 1982). This effect has been attributed to the more rapid conversion of $SO_2$ into $SO_4^{2-}$ in spring compared to winter, because of the higher solar radiation and the larger oxidant concentrations (Joranger *et al.*, 1980).

Apart from the seasonal trends in dry deposition of acidifying pollutants attributable to fluctuations in rates of emission, it should be remembered that, for deciduous trees, the absence of foliage also causes a marked drop in dry deposition rate (e.g. Höfken, 1983). This is also often discernible in throughfall for ions from maritime-derived aerosol inputs (see e.g. Miller 1985b).

Although much attention over recent years has focused upon the fact that atmospheric acidifying pollutants, particularly $SO_2$, may be transported over many hundreds or even thousands of kilometres from their original source, it should not be overlooked that there are rural and semi-rural areas where the extent of pollution varies dramatically with wind direction in relation to the bearing of a particular pollution source, perhaps within a few tens of kilometres (Ronneau *et al.*, 1978). Climatic influences on transformation rates may then become very important (Ronneau and Snappe-Jacob, 1978). Underwood *et al.* (1983), for example, found evidence for an increase in deposition of $H^+$, non-marine sulphate and nitrate in rural Nova Scotia along a north-east to south-west axis. This could be accounted for because 80% of the wet sulphate deposition over the four-year study period occurred in association with south-westerly, westerly or north-westerly winds.

## Interactions of acidic deposition with vegetation

The impact of vegetation, particularly tree growth, was considered briefly in Chapter 2 and again in the early part of this chapter, but only with reference to soil formation or possible changes in the hydrological pathways followed by drainage water. Interception of precipitation by vegetation may also considerably influence the solute composition of the water however. This section is concerned with the differences between the solute chemistry of throughfall and stemflow and incident precipitation. A great deal of work has been done on this aspect, both in the field and in the laboratory, and only the major relevant conclusions can be briefly considered here. More information may be found in the useful review by Irving (1983), which summarises effects observed prior to 1982 of acid rain for a range of crops, and the even more relevant review by Miller (1984b), which concentrates on the more significant (in the context of this monograph) effects of trees.

For convenience, acid deposition interactions with vegetation may be considered here under five broad headings:

  (i) Effects upon growth and subsequent indirect effects of impaired or improved growth. This heading includes ecophysiological effects, e.g. accumulation of S and/or N in needles, damage to needle waxes, changed susceptibility to fungal or insect attack, etc. (Huttunen, 1985).

  (ii) Effects upon cation leaching, including crown leaching from trees and cation exchange at the leaf surface.

  (iii) Effects of anion uptake, particularly in the case of nitrate.

  (iv) Effects upon leaching of organic species from vegetation.

  (v) Effects of the above-ground part of plants upon the $H^+$ flux to the rhizosphere.

It is appropriate to consider each of these topics briefly in turn

*Growth effects*

Possible effects of acidifying pollutants upon forest growth have been eloquently and concisely summarised by Abrahamsen (1983a). The pollutants may physically damage tissue and/or disturb the normal plant physiological processes. In the immediate vicinity of gross pollution sources, such as the Sudbury smelter in Ontario, the symptoms of $SO_2$ damage to vegetation are immediately apparent over a wide area (complete vegetation kill over 65 km$^2$ and destruction of all forest cover over an additional 225 km$^2$ (McLaughlin, 1985)). However, it is also possible to get reduced plant growth with no visible damage to tissue (Abrahamsen, 1983a). On the other hand, in areas where plants are deficient in N and/or S, deposition of the elements in a plant-available form may have a fertiliser effect (Abrahamsen, 1983a, 1984). Thus, except in grossly polluted areas, it is difficult to predict the effects of pollutants on growth, and each specific case must be considered separately. The complex interacting effects of fluctuations in growth rate have been considered elsewhere, and will not be considered again here. Pollutant effects must, of course, be considered in conjunction with climatic fluctuations. Small growth effects are probably of little consequence at most sites. Ecophysiological changes, such as modifications of the wax coatings on pine needles (Huttunen, 1985), are particularly interesting because of their possible effects upon susceptibility to cation leaching and upon the biodegradability of litter. The latter is currently under investigation in our laboratory in Aberdeen.

*Effects upon cation leaching*

The fact that cations may be leached from leaf surfaces by rainfall has been recognised for many years. In 1965, Mecklenburg *et al.* (1966) demonstrated, using plants containing $^{45}Ca$ absorbed through the roots, that calcium was leached from cation exchange sites within the foliage by ion exchange and diffusion processes. Early studies of the nature of substances leached from the above-ground parts of plants have been discussed in a comprehensive review by Tukey (1970). The possible implications of the phenomenon in the context of acid rain have been discussed more recently (Tukey, 1980).

In an experiment with simulated mists at pH 2.8, 4.3 and 5.7, Scherbatskoy and Klein (1983) demonstrated increased leaching of potassium and, to a lesser extent, calcium with increasing acidity. Their test crops were seedlings of yellow birch and white spruce. They also demonstrated that, with an increasing number of leachings, potassium was not readily resupplied to leaching sites. Very similar results were reported by Skiba *et al.* (1986) for 1.25-m pot-grown Sitka spruce trees. The latter study demonstrated that simulated rain acidified to pH 3.5 with sulphuric acid, compared to rain at pH 5.3, caused significant additional leaching of magnesium, potassium, calcium and manganese, and the sum of the additional cations leached (in µeq) was balanced by $H^+$ neutralisation, supporting the cation exchange mechanism hypothesis. More recently Skiba and Cresser (1986) have demonstrated that detached leaves, i.e. litter, behave in a very similar fashion. Leaching of magnesium and manganese from litter was particularly dramatically enhanced.

One of the objectives of the earlier work by Skiba *et al.* (1986) was to see if mist collected on vegetation over a prolonged period led to accumulation of acid on the surface, and a subsequent pulse of very acid throughfall when the next rain event occurred. In the event, the acid was neutralised to a large degree, but lowering the mist pH to values below 4 did not appear to cause any additional base cation leaching. An unexpected result of this misting experiment was that mist acidified with sulphuric acid to a specified pH often caused significantly more leaching than mist acidified to the same pH with nitric acid. An identical trend was observed for prolonged simulated rainfall experiments. The effect could not be fully explained in terms of nitric acid uptake.

Under field conditions it is much more difficult to discriminate between leaching, such as that described above, and wash-down in a shower of dry deposited material filtered out of the atmosphere on to the vegetation surface between storm events (Miller and Miller, 1980; Miller, 1985b). This filtering effect may be of great practical significance in the context of drainage water acidification, but not because of remobilised 'filtered' acid – most of this will probably be neutralised in the long contact time available. Where a tree collects high amounts of neutral salts between storms, these will be washed off fairly early in a subsequent storm (Skiba and Cresser, unpublished results). This high solute concentration may effectively lower the pH of water draining from surface horizons by displacing $H^+$ from cation exchange sites. Thus the higher the throughfall salt concentration, the lower the soil solution and drainage water pH is likely to be.

From the preceding discussion it might be anticipated that vegetation would always exert a neutralising effect upon acid precipitation, but this presupposes an ample and renewable supply of base cations on the leaf exchange sites. Miller (1984b), puzzled by the apparently often conflicting reports as to whether trees had an acidifying or a neutralising effect on rainwater, subdivided the available data from numerous studies around the world according to whether they were based upon temperate or tropical hardwood forests, conifer forests less than 60 years old or conifer forests more than 60 years old. Old conifers tended to acidify, whereas the other forests tended to neutralise (the hardwoods more effectively than the younger conifers). Miller also noted that effective neutralisation by deciduous species is confined to the summer period when they are in leaf. A similar seasonal trend is sometimes observed for conifers (Miller, 1983, 1985b). It seems possible that replenishment of base cations by root uptake from the acidified soils under mature trees is eventually inadequate to cope with the wet and dry deposited acid loading as the tree matures, and the tree is thus no longer capable of neutralising even the dry deposited acid, so some of the latter is remobilised, acidifying subsequent throughfall. Stemflow is consistently more acidic than incident precipitation (see e.g. Miller, 1985b). The acidification effect tends to be greater the more heavily an area is polluted. Stemflow delivers a small amount of water to a highly localised soil zone around the tree trunk.

At first it might appear that leaf cation exchange may, in many circumstances, significantly ameliorate the effects of acid rain before it contacts the soil. For example, neutralisation of *ca* 2000 moles of

H$^+$ ha$^{-1}$ a$^{-1}$ has been suggested for young Sitka spruce (Cresser *et al.*, 1986). However, it must be remembered that base cation uptake to replace leached cations effectively transfers the neutralised H$^+$ flux from the leaf to the rooting zone. The significance of acid generation in the rhizosphere is discussed later. Suffice it to say here that the buffering of acid rain by leaf tissue to higher pH values may be ecologically significant more because of the protection it affords to crops against damage (Craker and Bernstein, 1984) than because of its beneficial effects to surface waters.

*Anion uptake effects*

There is a growing body of evidence to suggest that nitrate uptake may be very significant in upland catchments susceptible to water acidification. Some of this evidence comes from analysis of seasonal trends in river-water composition such as that shown in Fig. 3.7 (Edwards *et al.*, 1984b). From very early to very late in the growing season very little

Fig. 3.7. Typical seasonal trend in river-water nitrate–N concentration at Glendye, Grampian Region.

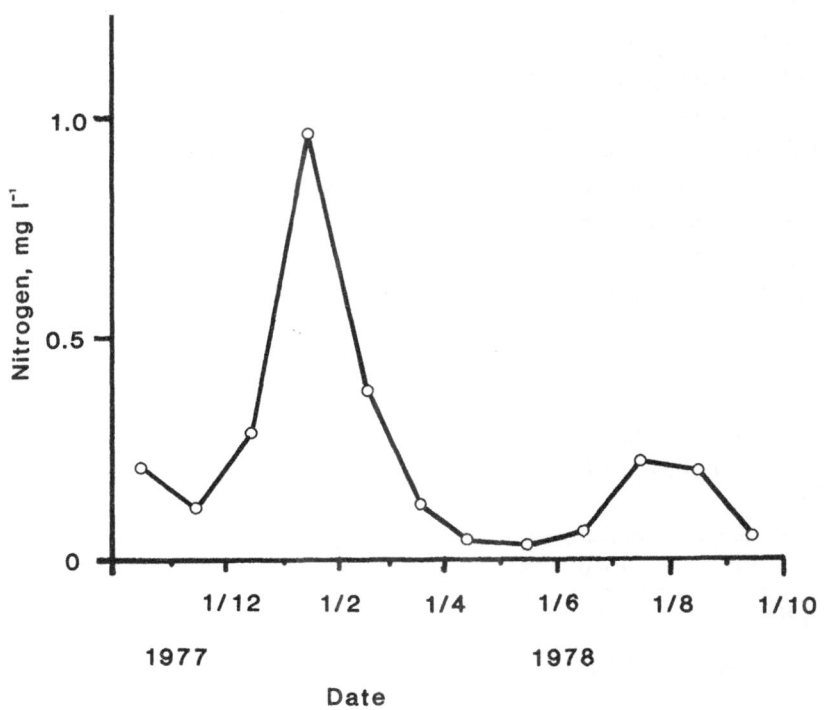

nitrate enters the river, showing the effect of retention of atmospheric nitrate inputs by the plant/soil system. More direct evidence comes from comparison of the incident rainfall and throughfall nitrate concentrations at moorland sites (Edwards *et al.*, 1985a). Both sphagnum and *Calluna vulgaris* were shown to be very efficient at removing nitrate from incident rain, so that nitrate in throughflow from surface organic horizons was negligible. Some typical results for throughflow are shown in Table 3.1. Nitrate uptake from acid rain has been observed for young Sitka spruce (Skiba *et al.*, 1986) and also for Sitka spruce litter in long-term experiments with simulated rain (Skiba and Cresser, 1986). Abrahamsen (1984) has observed $NO_3^-$ uptake from rain in Norwegian forests, and attributed the effect to the limited supply of nitrogen in Scandinavian forest soils.

The direct uptake of nitrate by plants is effectively a proton sink (van Breemen *et al.*, 1983). The reaction may be represented by the equation:

$$ROH + NO_3^- + H^+ \rightarrow RNH_2 + 2O_2$$

So long as the nitrogen remains in the plant, in plant litter or in soil organic matter, the $H^+$ input is effectively stored in the catchment area. Only when mineralisation and nitrification occurs, i.e. the reverse of the above reaction, is the $H^+$ and an associated highly mobile anion liberated with the possibility of mobilisation in drainage water.

The interesting question that arises from the above observations is: To what extent is nitrate uptake from rain dependent upon the underlying soil being nitrogen deficient? Work with sphagnum in heavily polluted areas suggests that sphagnum loses its ability to absorb precipitation nitrate under such conditions (John Lee, pers. comm.). Certainly, in field studies of tree throughfall, nitrate uptake is not always observed. More work is needed to provide a definitive answer to this question.

Table 3.1. *Nitrate–N ($\mu g\ ml^{-1}$) in rain and throughflow from a 0–100 mm surface horizon at Peatfold, Glenbuchat, in the Grampian Region, Scotland*

| Date | | 9 Sept. 1982 | 16 Sept. 1982 | 13 May 1983 | 15 Sept. 1983 |
|---|---|---|---|---|---|
| $NO_3^-$–N | Rain | 0.09 | 0.39 | 0.73 | 0.40 |
| | Throughflow | 0.01 | 0.08 | 0.00 | 0.00 |

Uptake of sulphate by plants appears to be of little consequence. Both organic peat and mineral soils may retain sulphate, however. This phenomenon and its significance are discussed later in this chapter.

*Effect of acid precipitation on leaching of organic compounds*
Much less work has been done on the influence of precipitation acidity upon the leaching of organic species from plants than on the leaching of inorganic species. This is probably due in part to the difficulty of chemical analysis of the throughfall for organics present at very low concentrations. However, Scherbatskoy and Klein (1983) managed to show that acidified mist increased the leaching of amino acids and carbohydrate from young birch and spruce. The pH of the mist they used seems very low at 2.8, but it should be remembered that suspended water droplets (mist, fog, etc.) may be much more acidic than typical rain at a given site (Dollard *et al.*, 1983).

The soluble organic components in throughfall have been shown to considerably stimulate soil microbial activity as reflected by respiration rate (Cresser *et al.*, 1986). It is possible, therefore, that increased foliar leaching of carbohydrate and amino acids could more than compensate for any suppression of microbiological activity induced by the effect of the acid precipitation.

*Gaseous input effects*
Most of the discussion so far has been concentrated upon the significance of acidifying pollutants in precipitation, and dry deposited salts. However, the significance of gaseous inputs, particularly of sulphur dioxide, and mist, must also be considered before we consider what happens in the root zone. Hällgren *et al.* (1982) studied the net fluxes of $SO_2$, water vapour and carbon dioxide to one-year-old and current shoots of Scots pine in the field using fumigation in temperature-controlled assimilation chambers. They investigated the $SO_2$ deposition flux at atmospheric concentrations of 40 to 250 $\mu$g m$^{-3}$ (15–95 ppb). They found marked diurnal variation in the uptake of the gas, deposition velocity being three times less at night than during the day during the summer. Part of the difference was attributed to light-dependent re-emission of sulphur compounds by the needles. Previously it had been generally assumed that the deposition flux depended upon stomatal opening and that the sorption was irreversible. The more complex behaviour outlined above should be borne in mind when considering possible effects of the gas upon cation leaching. The assimi-

lation chamber technique suffers from the drawback that the shoot is protected from precipitation, so concurrent information on leachate chemistry is not available. In growth chamber experiments interest has centred around possible physiological damage to the plant rather than around the effects reflected in throughfall. This is unfortunate, because the latter are of more interest in the present context unless tree growth is severely retarded.

In heavily polluted areas, especially afforested areas, dry deposition of gaseous $SO_2$ may be very substantial. Paces (1985), for example, has estimated such inputs to an industrially damaged forest in Czechoslovakia to be as high as $88.6 \, \text{kg ha}^{-1} \, \text{a}^{-1}$. Such a flux oxidised to sulphuric acid would yield as much $H^+$ as 2400 mm of rain at pH 4. The sulphate output of $96 \, \text{kg ha}^{-1} \, \text{a}^{-1}$ in runoff from the catchment was accompanied by correspondingly high base cation leaching. Paces (1985) noted that the high $SO_2$ gradient from the unpolluted to the polluted site correlated with dieback of coniferous trees and acceleration of weathering and erosion, accompanied by increased output of nutrients. For reasons discussed elsewhere, a high neutral salt output may lower drainage water pH considerably. Prolonged neutral salt removal could stress the ecosystem and lead to lowering of soil base saturation and pH. Paces (1985) also commented that the twenty-fold increase in nitrate in runoff from the damaged forest compared to a relatively pollution-free forest could not be accounted for in terms of $NO_x$ pollution input. Instead he attributed the higher nitrate flux to decreased denitrification capacity in the damaged forest, arguing that this agreed with the observed increased nitrate in runoff from a similar forest in New Hampshire which had been clear-cut (Likens *et al.*, 1977). To us, it seems probable that reduced nitrate uptake following nitrification, even in the acid forest soil, could contribute to the leaching losses of nitrate. In Paces' (1985) study, the nitrate flux in runoff from the damaged forest almost exactly balanced the total inorganic nitrogen (nitrate-N plus ammonium-N) in rain. Whatever the mechanism, the important point here is the mobilisation of nitrate and associated base cations out of the ecosystem, leading to soil acidification and potential further water acidification.

There can be little doubt that gross $SO_2$ pollution may lead to substantial soil acidification. Some of the more dramatic evidence for this has been discussed by Tomlinson (1983a). The graphs in Fig. 3.8 are based upon the results of Hutchinson and Whitby (1977) cited by Tomlinson. They show how the pH and soluble aluminium concentrations at the soil surface and at a soil depth of 5 cm varied with distance from the Coniston smelter in Ontario. These results really require no further comment here, apart perhaps from the need to stress the acuteness of the problem – deposition pH was 2.85 at 1.6 km, and still only 3.72 at 13.5 km. Evidence for reduction in soil pH as a result of more typical pollutant inputs is obviously less impressive; it is also less available. Some effect clearly may occur when susceptible soils are stressed, but it is very difficult to discriminate experimentally between gaseous input effects and wet and other dry deposition effects.

Before leaving the subject of gaseous acidifying pollutant effects, it is important to remember the possibility of synergistic effects of $SO_2$ with other air pollutants, especially ozone and oxides of nitrogen (Abrahamsen, 1983a). These effects are perhaps more relevant to forest and agricultural crop damage, however, and need not be considered further here.

Discussion has concentrated upon $SO_2$ effects so far. This partly reflects its relatively greater importance in most situations, and partly our more restricted knowledge of $NO_x$ effects. Effects of ammonia have been well documented, especially by van Breemen *et al.* (1982, 1983, 1984), and discussed elsewhere in the present text.

Fig. 3.8. Effect of distance from the Coniston smelter (km) upon pH (left) and soluble aluminium (right) in soil at the surface (open circles) and at 5 cm (filled circles). Based upon data cited by Tomlinson (1983a).

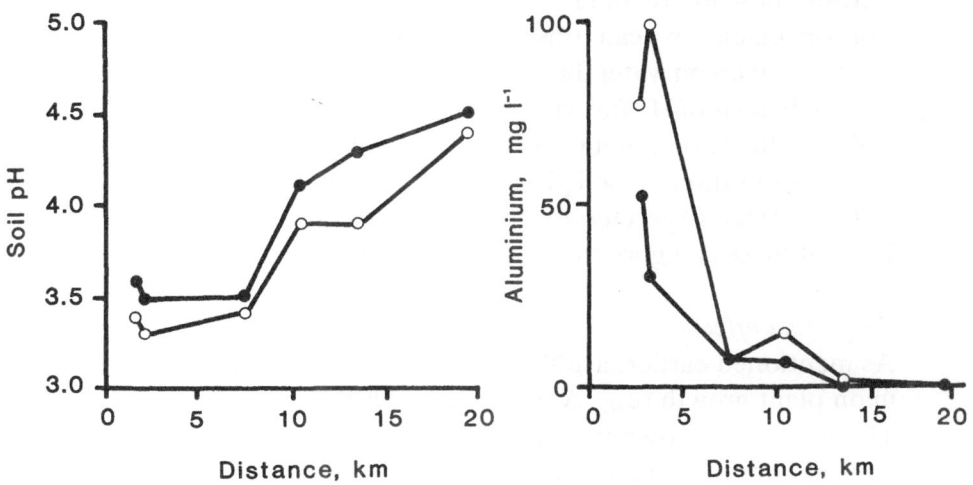

*Mist effects*

In spite of the high solute concentration and low pH of mist (Dollard *et al.*, 1983), very little field information is available upon the effects of acidic mists on the soil/plant/water system. This is primarily because of the difficulty in isolating rain, mist and dry deposition effects from each other when the boundaries of the categories are anyway indistinct. From the work of Craker and Bernstein (1984) it might be expected that the longer contact time of mist with foliage compared to rain would lead to greater opportunity for ion exchange and therefore more effective neutralisation of $H^+$ and base cation leaching. Continuous exposure to simulated acid mist (4 to 6 hours) certainly leads to substantial base cation leaching in the throughfall (Scherbatskoy and Klein, 1983; Wood and Bormann, 1975) and some neutralisation (Scherbatskoy and Klein, 1983). Intermittent misting apparently leads to a more complete neutralisation, as might be expected (Cresser *et al.*, 1986; Skiba *et al.*, 1986), but without the expected proportionate base cation leaching when eventually the foliage becomes wet enough for throughfall to occur. This is possibly the result of foliar uptake. Aside from this effect however, which would presumably also occur from rain retained by wetted vegetation, there appears to be little difference between rain and mist apart from the contact-time effect.

## The significance of vegetation effects in relation to acidifying pollutants

Having considered the nature of the interactions of acidifying pollutants with vegetation, it now seems appropriate to assess how significant these interactions are in the context of freshwater acidification. For convenience we can consider the effects under four main categories:

  (i) Effects on water flux via effects on growth, especially of trees.
 (ii) Effects of $H^+$ flux via amount of $H^+$ 'trapped'.
(iii) Effects on amount of $SO_4^{2-}$ trapped, and hence, via sulphate saturation, on soil pH and acidification.
 (iv) Effects of accelerated base cation leaching from foliage.

Each of these categories is briefly assessed in turn.

*Water flux effects*

As mentioned earlier, acidifying pollutants may exert a fertiliser effect upon plant growth (e.g. Abrahamsen, 1983b) or they may suppress it. The impact of either improved or impaired growth upon water flux depends upon a complex range of interacting factors (Cresser *et al.*,

1986). These have already been considered when the significance of changes in land use was discussed earlier in this chapter. Unless the effect upon growth is gross, water flux effects upon surface water acidity are likely to be small. If, however, the lowering of base saturation of soil arising from atmospheric acidifying pollutants is substantial, to the extent that root growth patterns are appreciably modified, then changes in soil structure and hence the hydrological pathway of drainage water could occur, and these could be significant.

*Effect of $H^+$ flux and transfer of $H^+$ to the root zone*
Conceptually the simplest and most obvious effect of vegetation is probably that substantial growth, such as the canopy of well-developed forest, may trap a significantly greater amount of extra acidifying pollutants than say the vegetation on a moorland or agricultural site. When the additional $H^+$ flux is the result of more efficient trapping of dry deposition, including gaseous inputs, the long contact time between pollutant and vegetation means that much of the $H^+$ will be neutralised by cation exchange. However, it must be assumed that, if the foliar base cation content is maintained, foliar leaching is balanced by root uptake. The latter in turn results in the transfer of an equivalent $H^+$ flux to the rhizosphere.

Whether or not the additional $H^+$ flux in the root zone will significantly lower the soil pH or not depends primarily on the capacity of the soil mineral assemblage to release base cations by weathering at a sufficient rate to maintain the soil base saturation. If it does not, then the base saturation of the soil, and hence its pH, will fall.

Nilsson *et al.* (1982) and Miller (1985a) suggested that root-generated acidity varied only gradually, occurred in intimate contact with the soil, and did not involve anion movement. Thus, compared to acidic precipitation episodes, they concluded that it was less likely to lead to streamwater acidification. It is difficult to see how this hypothesis can apply in areas where substantial amounts of neutral salts are deposited however. Ion exchange equilibria are very rapid, and therefore lowering of soil pH is likely to lead to a lowering of drainage water pH as a result of cation exchange. If acid deposition causes more leaching of base cations from foliage and soil, and thus leads to reduced soil pH in the root zone, this would also lower the pH of water draining from the reduced pH soil horizon. The acidification of the root zone at many sites will only be reflected by water acidification during storms, when significant lateral flow may occur. Trees may trap more acidifying pollutants and also more neutral salts. Thus, especially when base cations stored in the trees are considered too, trees increase soil acidification and the probability of drainage water acidification in the long term. Where they substantially increase the $SO_2$ and sulphate deposition fluxes, they may also lead to sulphate saturation of anion exchange sites and thus to acidification of the soil at depth, as discussed in the next section. Miller (1985b) commented that some of the foliage-derived base cations in throughfall could come into contact with the relevant soil cationic exchange sites, displacing root-derived $H^+$ into drainage water and thus yielding a closed cycle of base cations. However it must be remembered that rainfall acidification reduces the retention of base cations in near-surface horizons (see e.g. Cresser *et al.*, 1986).

*Sulphate saturation effects*

The importance of anion exchange reactions involving sulphate in soil was considered briefly in Chapter 2, and the effect of trees trapping more sulphate earlier in this chapter. It is appropriate to consider at this point the possible impact of high levels of pollutant-derived sulphate upon the working of the sulphate cycle. In a typical upland acid moorland soil, sulphate leaching through mineral soil is limited by adsorption by iron and aluminium oxyhydroxides in the profile. As described in Chapter 2, this serves to minimise base cation leaching (i.e. soil acidification in this context). The question that must be asked is: Why do the oxyhydroxides not become saturated with sulphate? The answer appears to be that sulphate, iron and aluminium are all being continuously recycled and cycled through the soil/plant/water system. Thus new oxyhydroxide surfaces are being continuously created, providing new sites for sulphate anion exchange. Provided new site production keeps pace with sulphate inputs, the anion exchange process will continue to stabilise soil pH at depth. The importance of sulphate adsorption can be clearly seen in the work of Lee (1985). Lee subjected forest soil in large, deep lysimeters under sugar maple or red alder to 'rain' acidified to pH 4.0, 3.5 or 3.0 for 3½ years. His results showed sulphate accumulation above 45 cm, and only the top 15 cm of soil showed a significant pH reduction. Tomlinson (1983a) summarised work in the Adirondacks in New York State which demonstrated sulphate and calcium accumulation in the B horizon (i.e. concentrations were lower in leachate from the B horizon than in leachate from the A horizon). Matzner (1985) has attributed sulphate retention prior to 1974 at Soller to jurbanite $(AlOHSO_4)$ formation rather than sulphate anion exchange.

Johnson and Henderson (1979) pointed out that, for their site in eastern Tennessee, further sulphate inputs might be adsorbed to some extent by the surface horizons, but once their adsorptive capacity was exceeded sulphate could percolate relatively freely through the soil. High pollutant sulphate loading may rapidly disrupt the delicate balance between oxyhydroxide precipitation rate and sulphate adsorption rate, leading to sulphate saturation and concurrent leaching of sulphate and base cations in drainage water (Tomlinson, 1983a, b). This breakthrough effect is often clearly visible in lysimeter studies (see e.g. Morrison, 1983). Fall in drainage water pH tends to lag behind the base cation and sulphate breakthrough because of the time taken to significantly lower the soil base saturation (see e.g. Morrison, 1983).

Cresser *et al.* (1986) have examined the sulphate adsorption properties of upland acidic moorland soils in north-east Scotland. Their study was prompted by the observation that a catchment in which the soils were derived from granite (Glendye) had virtually no capacity to retain sulphate, whereas another catchment, Peatfold, with soils evolved from the more basic quartz–biotite–norite but otherwise very similar, adsorbed sulphate. It was found that reduction of simulated rainfall pH to 3.5 with sulphuric acid still left scope for adsorption of sulphate for up to 12 weeks in the B horizon from a mid-slope soil pit at Peatfold (Fig. 3.9). The Glendye-B horizon soil was sulphate-saturated by this 'rain' within just a few days. Sulphate breakthrough was associated with

Fig. 3.9. Patterns of sulphate adsorption or solubilisation (expressed relative to input in rain) from 0–50 (open squares), 0–150 (open circles) and 0–250 (open triangles) mm depth when lysimeters were subjected to 'rain' acidified to pH 3.5 with sulphuric acid.

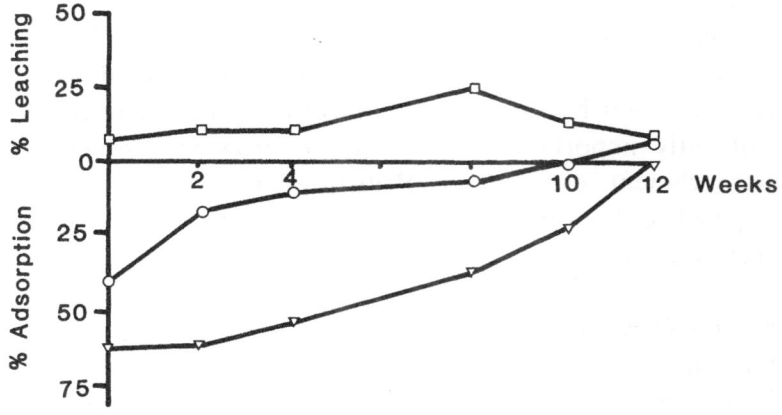

increased base cation leaching in every case. Examination of soils along altitudinal transects in the two catchments showed that sulphate saturation was greatest on the lower slopes, presumably because of inputs in lateral throughflow.

Krug and Frink (1983) suggested that the hypothesis that increased deposition of acid sulphate caused equivalent leaching soil acidification was theoretically unsound for acid soils. They postulated that the sulphuric acid input lowered the solubility of organic acid anions with negligible pH change, and simple cation exchange was unimportant so that acid rain did not increase the base cation leaching loss from very acid soils. In our experience cation exchange in surface horizons cannot be ignored, but displaced base cations are readsorbed at depth alongside sulphate. Sulphate saturation does not always occur, even in a moderately heavily polluted area, and therefore quantitative substitution of a base cation flux in drainage water to balance the $H^+$ flux in rain would not necessarily be expected. Indeed a 1:1 equivalence is always most improbable, because some sulphate adsorption is nearly always possible in mineral soils.

A complicating factor when considering sulphate adsorption in different soils is the possible impact of dissolved organic matter adsorption on to the relevant oxyhydroxide surfaces, which reputedly blocks them against sulphate adsorption (Krug and Frink, 1983). On this basis, because rainfall acidity lowers the solubility of organic matter in the surface horizon, it might be anticipated that more sulphate adsorption would occur as a result of acid precipitation.

It has already been mentioned that establishment of trees may lead to substantial extra inputs of wet and particularly dry deposition to a catchment. Thus sulphate saturation is more likely in a forested site than at a moorland site. Recent work by us and some of our colleagues, as yet unpublished, has shown that bracken (*Pteridium aquilinum*) is also a highly efficient collector of dry deposition, and bracken throughfall may contain ten times as much sulphate as *Calluna* throughfall. This observation is currently important because of the growing problem of bracken encroachment, especially in British uplands (Taylor, 1985). Skeffington (1983) did not find similarly high concentrations of solute under bracken, however.

*Foliar base cation leaching effects*
The possible importance of the salt effect, i.e. the lowering of soil solution pH as a result of the presence of neutral soluble salt, has been

discussed earlier in this chapter in another context. Its significance was succinctly explained by Wiklander (1975) early on in the acid rain debate. There are many published examples which illustrate crown leaching effects and the collection efficiency of different plant species for salts (e.g. Miller and Miller, 1980; Matzner, 1983; Skeffington, 1983; Ulrich, 1983a, c). Harriman and Morrison (1981) showed that the greater deposition to forest was reflected in drainage water chemistry, although some allowance had to be made for the greater evapotranspiration from the forested catchment. The presence of additional soluble salts may significantly lower the pH of the soil solution and hence of drainage water. Acidity of rain increases the base cation content of throughfall compared to that of the incident precipitation. This may further lower the drainage water pH. Rosenqvist (1978) showed that typical simulated coastal and inland rainwaters leaching acidic humus consistently gave drainage water with a pH value 0.2 units lower for the coastal 'rain'. This gives some indication of the order of magnitude of the change that might be expected as a consequence of the salt effect.

It is worth bearing the salt effect in mind when pH measurements are made to look for trends in pH of soils over 20 to 40 years or more. Some workers use dilute calcium chloride solution to make a thick slurry to measure pH. Many others simply add water for pH measurement when comparing soils (e.g. Thompson and Loveland, 1985). In the latter instance, preceding rainfall or throughfall solute composition and concentration may be very significant, as may be the presence or absence of dry weather for a prolonged period prior to sampling. The effect is particularly important when the pH change sought is small.

Foliar effects on leaching and dry deposition might be expected to exhibit seasonal trends, and this is indeed the case, especially for deciduous trees (Hoffman *et al.*, 1980).

## Acid deposition–soil interactions

Most of what needs to be said about acid deposition effects upon soils has been said already in Chapter 2 and the earlier part of this chapter, with the exception of acid deposition effects upon microbial activity. A brief summary of the important points could be useful here, however.

Water draining through soil equilibrates rapidly with ion exchange sites. Thus the water rapidly assumes a pH dominated by soil pH, but influenced by the soluble salts in the water. If the rainfall is acidified by the presence of pollutants, the base saturation of the soil will be temporarily lowered slightly. Mineral weathering between storms may replenish base cations leached from cation exchange sites during the storm. If it does this at an adequate rate, the soil base saturation does not fall. If it cannot, base saturation will fall to a level which depends upon residual mineral weathering capacity and precipitation chemistry. Thus, in the long term, acid deposition lowers the surface soil pH of susceptible soils, and consequently may lower drainage water pH.

In upland areas, depending upon the soil fertility, sulphate and chloride tend to be leached down the soil profile and nitrate tends to be taken up by the surface horizon biomass. Sulphate tends to be adsorbed at depth, leading to concurrent adsorption of base cations leached from higher horizons. Unless the sulphate adsorption sites at depth are saturated, chloride tends to dominate the anions in the drainage water. Sulphate inputs may be substantial from lateral flow however, unless the ion is retained by anaerobic peats adjacent to streams (Brown, 1985a, b; Brown and MacQueen, 1985). Heavy pollution by sulphur species increases the probability of sulphate saturation and the associated increased base cation leaching. In the long term, where mineral weathering is well advanced, the result may be soil acidification at depth, and, ultimately, drainage water acidification.

From the preceding summary of acid deposition/soil interactions, it might be assumed that immediate direct effects of acid rain upon soil microbial activity are improbable, because the soil solution pH is very rapidly dominated by soil pH rather than precipitation pH. Only when the 'rain' added to soil is very acid ($\leqslant$pH 3) is a direct effect generally observed (Cresser et al., 1986). Killham et al. (1983) simulated rain at pH 2, 3, 4 and 5.6 falling upon a Sierran forest soil at pH 6.4 over 12 weeks. As might be expected, effects were greatest in the top 1 cm of soil, but only the pH 2 input causes inhibition of both respiration and enzyme activities. Indeed, rain at pH 3 and 4 sometimes exerted a stimulatory effect. Johnson and Todd (1984) examined the effect of sulphuric acid and nitric acid loadings at two times and ten times the natural levels over one year on an acid forest soil (pH 4.6–5.0). The soil was initially derived from dolomite at the Walker Branch Watershed near Oak Ridge, Tennessee. The treatments had no statistically significant lasting effect on soil ammonium, nitrate, extractable phosphorus, extractable aluminium, carbon dioxide evolution, mineralisable nitrogen or nitrification potential. Langkramer and Lettl (1982) studied changes in soil microbiology with distance from an emission source in western Bohemia. The actual and potential respiration of soils and the decomposition of cellulose, the oxidation or thiosulphate and, particularly, ammonification were inhibited. Nitrification was unaffected, but oxidation of elemental sulphur was slightly stimulated.

The few examples cited above have been selected because they typify the findings of soil microbiological studies of acid rain effects. Short-term studies, i.e. $<ca$ 12 months, based upon simulated rain, only tend to show effects when the rain is very acid (pH 3 or lower), and then only in the top centimetre or so of soil. Their results must be interpreted with care unless the 'rain' has been added at a realistic rate (Killham and Firestone, 1982). Field studies at sites which are (and have been for several years) subjected to heavy acidifying pollutants loadings do suggest modification to microbial activity. The long-term effect may be due to reduction in surface soil pH over a long period. Some subtle indirect effects also occur. For example, Skiba and Cresser (1986) found that leachate from acid-rain-treated spruce litter suppressed the growth of bacterial cultures isolated from a forest soil, and the effect could not be explained in terms of pH or increased manganese solubility. Tyler (1983) has suggested that acid deposition could increase the toxic effects of heavy metals in soils.

Changes in microbial activity could be important in the present context for a number of reasons. Reduced microbial respiration could reduce the $CO_2$-buffering effect discussed in Chapter 2. Reduced decomposition rates could lead to accumulation of increasing thicknesses of acidic litter at the surface which could become important during storms if they contributed to a lateral flow effect. Nitrogen and sulphur cycle effects could also be significant. However it must be stressed that little firm evidence has been found for significant effects in the context of freshwater acidification at the time of writing.

One other aspect of the microbiological behaviour of soil sulphur must be considered. Work by Brown (1985a) has shown that, in anaerobic peat, sulphate is reduced to hydrogen sulphide, probably by dissimilatory sulphate-reducing bacteria. It was estimated that sulphur had been accumulating in the Tillingbourne peat at a rate of 4.8 to $6\ \mathrm{g\,m^{-2}\,a^{-1}}$ (Brown, 1985b). Brown and MacQueen (1985) commented that, since the water flowing through mires also generally transports dissolved oxygen, it seemed most probable that sulphate reached the reducing sites by diffusion through aerobic or microaerophilic zones in which the other micro-organisms consumed the oxygen to create the low redox ($E_h < 100\ \mathrm{mV}$) conditions required. The importance of the hydrogen sulphide accumulation is most likely to manifest itself if for some reason the aeration of the peat improves to give oxidising conditions.

$$H_2S + 2O_2 \rightleftharpoons SO_4^{2-} + 2H^+$$

The oxidation could therefore produce a flush of sulphuric acid. Christopherson *et al.* (1984) included the reduction or adsorption of water-soluble sulphate in their model for sulphate behaviour in the Storgama catchment in Norway.

*Rates of soil acidification*
Nilsson (1983) reviewed the evidence available about causes of soil acidification up to 1982 and concluded that there was no unequivocal evidence for soil acidification 'mainly caused by atmospheric deposition'. He claimed that soil pH falls detected after a ten-year gap from initial measurements could have been natural acidification due to tree growth. This is in marked contrast to the earlier conclusions of Overrein (1972), who suggested that the acidity of a particular soil was regulated by four factors, the concentration excess of acid in precipitation, the susceptibility to leaching of nutrient elements available, the amount of water draining through the profile, and the buffer capacity of the soil. He therefore inferred that soils gradually acidified, leading to a gradual acidification of groundwater. We know now that the situation is more complicated than these simple concepts suggest. However, most of the evidence for the hypothesis that acid deposition acidifies soils remains rather circumstantial, being largely based upon extrapolation from lysimeter studies. Sometimes the evidence from such investigations is strong, for example the study of soil acidification in the vicinity of a coking works by Killham and Wainwright (1984). Lysimeters of unpolluted soil placed at various distances from the works showed marked falls in surface soil pH within three years in the highly polluted zones. Evidence for longer-term reduction in soil pH as a result of prolonged exposure to less-acid rain (e.g. pH 4.2 to 4.6) is less forthcoming.

Bearing in mind the circumstantial evidence and the rapid and substantial degree of soil acidification which may occur under trees (e.g. Skeffington, 1983; Ulrich, 1983b), it is hard to accept that acidic precipitation does not, in some circumstances, lower surface soil pH for a stressed ecosystem over a few decades. Where sulphate saturation of anion exchange sites has occurred, pollutant sulphate inputs should contribute to accelerated soil acidification to depth. Ulrich (1983a) regards such pollutant effects as forcing forest ecosystems from a stability range, buffered by recycling, to a transitional destabilisation phase. Van Breemen *et al.* (1983) calculated individual components of the $H^+$ budgets for a range of soils and watersheds. They showed that acid deposition sometimes may be a dominant input factor in some ecosystems. However, quantitative treatment of rates of soil pH fall has not yet been successfully achieved with respect to acid rain inputs. For calcareous soils, Breeuwsma and de Vries (1984) have concluded that natural soil acidification is much more important than the effects of anthropogenic-derived $H^+$ inputs.

## Direct acidification effects

Mechanisms by which water from precipitation may enter streams and, more especially, lakes with little change have been considered in Chapter 2. To recap briefly, direct entry of rain and snow may be significant for large, shallow lakes. High amounts of outcropping rock may significantly reduce the extent to which precipitation chemistry is modified. Blown snow may also be significant for large, shallow lakes. Where the precipitation contains excess acid, these processes may be regarded as direct acidification effects. They need not be considered further here.

## Acid rain, changes in land use or natural acidification?

It should be clear by now that there is rarely a simple, single answer to the question: What is the dominant cause of water acidification at this particular site? For some terrain, particularly steep slopes with extensive rock outcrops and shallow soils derived from susceptible materials, the drainage water should be naturally acidic during storms and snowmelt, but acid deposition can undoubtedly further aggravate the situation. Forest growth on a susceptible site may significantly lower soil pH and may trap more pollutants and more neutral salts. It may thus conceivably eventually cause water acidification. On the other hand, forest clearance may lead to iron pan formation under moorland vegetation, and thus, over a century or more, to water acidification by changing catchment hydrology. In every case climatic trends must also be taken into account. Superimposed upon this general plan however comes a wide range of local effects, including ploughing, liming and fertiliser use, erosion, growth effects, microbiological effects, soil structure effects, presence of impermeable horizons, sulphate saturation effects, etc. At present, then, it appears that each site must be considered individually.

## References

Abrahamsen, G. (1983a) Effects of long-range transported air pollutants on forest – summary document. In *Ecological Effects of Acid Deposition*, National Swedish Environment Protection Board, Report PM1636, 191–7.

Abrahamsen, G. (1983b) Sulphur pollution: Ca, Mg and Al in soil and soil water and possible effects on forest trees. In *Effects of Accumulation of Air Pollutants in Forest Ecosystems*, Ulrich, B. and Pankrath, J., eds, D. Reidel Publishing Co., 207–18.

Abrahamsen, G. (1984) Effects of acidic deposition on forest soil and vegetation. *Philosophical Transactions of the Royal Society of London*, **305**, 369–82.

Barrett, C. F., Fowler, D., Irwin, J. G., Kalland, A. S., Martin, A., Scriven, R. A. and Tuck, A. F. (1982) *Acidity of Rainfall in the United Kingdom – a Preliminary Report*, Warren Spring Laboratory, June 1982, report prepared at the request of the Department of the Environment.

Battarbee, R. W., Flower, R. J., Stevenson, A. C. and Rippey, B. (1985) Lake acidification in Galloway: a palaeoecological test of competing hypotheses. *Nature* (London), **314**, 350–2.

Booty, W. G. and Kramer, J. R. (1984) Sensitivity analysis of a watershed acidification model. *Philosophical Transactions of the Royal Society of London*, **B305**, 441–9.

Breeuwsma, A. and de Vries, W. (1984) The relative importance of natural production of $H^+$ in soil acidification. *Netherlands Journal of Agricultural Science*, **32**, 161–3.

Brimblecombe, P. and Stedman, D. H. (1982) Historical evidence for a dramatic increase in the nitrate component of acid rain. *Nature* (London), **298**, 460–2.

Brown, K. A. (1985a) *Formation of Organic Sulphur in Anaerobic Peat*, Central Electricity Generating Board Report No. TPRD/L/2886/N85, Leatherhead, Surrey.

Brown, K. A. (1985b) Sulphur distribution and metabolism in waterlogged peat. *Soil Biology and Biochemistry*, **17**, 39–45.

Brown, K. A. and MacQueen, J. F. (1985) Sulphate uptake from surface water by peat. *Soil Biology and Biochemistry*, **17**, 411–20.

Chalmers, A. G. (1985) Review of information on lime loss and changes in soil pH gained from ADAS experiments. *Soil Use and Management*, **1**, 17–19.

Christophersen, N., Rustad, S. and Seip, H. M. (1984) Modelling streamwater chemistry with snowmelt. *Philosophical Transactions of the Royal Society of London*, **B305**, 427–39.

Craker, L. E. and Bernstein, D. (1984) Buffering of acid rain by leaf tissue of selected crop plants. *Environmental Pollution (A)*, **36**, 375–81.

Cresser, M. S., Edwards, A. C., Ingram, S., Skiba, U. and Peirson-Smith, T. (1986) Soil-acid deposition interactions and their possible effects on geochemical weathering rates in British uplands. *Journal of the Geological Society, London*, **143**, 649–58.

Darling, F. F. and Boyd, J. M. (1964) *The Highlands and Islands*, Collins, London.

Department of the Environment Central Unit on Environmental Pollution (1976)

*Effects of Airborne Sulphur Compounds on Forests and Freshwaters*, a report of the Discussion Group of the Department of the Environment and the Natural Environment Research Council, Pollution Paper No. 7, HMSO, London

de Vries, W. and Breeuwsma, A. (1984) Causes of soil acidification. *Netherlands Journal of Agricultural Science*, **32**, 159–61.

Dimbleby, G. M. (1962) *The Development of British Heathlands and their Soils*, Clarendon Press, Oxford.

Dollard, G. I., Unsworth, M. H. and Harve, M. J. (1983) Pollution transfer in upland regions by occult precipitation. *Nature* (London), **302**, 241–3.

Edwards, A., Creasey, J., Reid, M., Cresser, M. and MacLeod, D. (1984a) Preliminary results of a comparison of weathering rates in two contrasting catchments in North-east Scotland. *Physio-Géo*, No. 9, 117–27.

Edwards, A. C., Creasey, J. and Cresser, M. S. (1984b) The conditions and frequency of sampling for elucidation of transport mechanisms and element budgets in upland drainage basins. *Hydrochemical Balances of Freshwater Systems*, IAHS Publication No. 150, Oxford, 187–202.

Edwards, A. C., Creasey, J. and Cresser, M. S. (1985a) Factors influencing nitrogen inputs and outputs in two Scottish upland catchments. *Soil Use and Management*, **1**, 83–7.

Edwards, A. C., Creasey, J., Skiba, U., Peirson-Smith, T. and Cresser, M. S. (1985b) Long-term rates of acidification of UK upland acidic soils. *Soil Use and Management*, **1**, 61–5.

Edwards, A. C., Creasey, J. and Cresser, M. S. (1986) Soil freezing effects on upland stream solute chemistry. *Water Research*, **20**, 831–4.

FitzPatrick, E. A., Mackie, L. A. and Mullins, C. E. (1985) The use of plaster of Paris in the study of soil structure. *Soil Use and Management*, **1**, 70–2.

Fowler, D., Cape, J. N. and Leith, I. D. (1985) Acid inputs from the atmosphere in the United Kingdom. *Soil Use and Management*, **1**, 3–5.

Franks, J. (1983) Acid rain. *Chemistry in Britain*, June, 504–9.

Gasser, J. K. R. (1985) Processes causing loss of calcium from agricultural soils. *Soil Use and Management*, **1**, 14–17.

Gimingham, C. H. (1964) Dwarf-shrub heaths. In *The Vegetation of Scotland*, Burnett, J. H., ed., Oliver and Boyd, Edinburgh, 232–88 (chapter 7).

Gimingham, C. H. (1972) *Ecology of Heathlands*, Chapman and Hall, London.

Gimingham, C. H. (1981) Conservation: European heathlands. In *Heathlands and Related Shrublands of the World, Vol. B, Analytical Studies*, Specht, R. L., ed., Elsevier, Amsterdam, 251–3 (chapter 26).

Goldstein, R. A., Gherini, S. A., Chen, C. W., Mok, L. and Hudson, R. J. M. (1984) Integrated acidification study (ILWAS): a mechanistic ecosystem analysis. *Philosophical Transactions of the Royal Society of London*, **B305**, 409–25.

Goulding, K. W. T. and Poulton, P. R. (1985) Acid deposition at Rothamsted, Saxmundham and Woburn, 1969–83. *Soil Use and Management*, **1**, 6–8.

Granat, L. (1983) Measurements of surface resistance during dry deposition of $SO_2$ to wet and dry coniferous forest. In *Effects of Accumulation of Air Pollutants in Forest Ecosystems*, Ulrich, B. and Pankrath, J., eds, D. Reidel Publishing Co., 83–9.

Granat, L., Rodhe, H. and Hallberg, R. O. (1976) The global sulphur cycle. In *Nitrogen, Phosphorus and Sulphur – Global Cycles, SCOPE Report 7, Ecological Bulletin (Stockholm)*, **22**, 89–134.

Grubb, P. J., Green, H. E. and Merrifield, C. J. (1969) The ecology of chalkheath: its relevance to the calcicole–calcifuge and soil acidification problems. *Journal of Ecology*, **57**, 175–212.

Hällgren, J. E., Linder, S., Richter, A., Troeng, E. and Granat, L. (1982) Uptake of $SO_2$ in shoots of Scots pine: field measurements of net flux of sulphur in relation to stomatal conductance. *Plant, Cell and Environment*, **5**, 75–83.

Harriman, R. and Morrison, B. R. S. (1981) Forestry, fisheries and acid rain in Scotland. *Scottish Forestry*, 89–95.

Hoffman, W. A., Lindberg, S. E. and Turner, R. R. (1980) Precipitation acidity: the role of the forest canopy in acid exchange. *Journal of Environmental Quality*, **9**, 95–100.

Höfken, K. D. (1983) Input of acidifiers and heavy metals to a German forest area due to dry and wet deposition. In *Effects of Accumulation of Air Pollutants in Forest Ecosystems*, Ulrich, B. and Pankrath, J., eds, D. Reidel Publishing Co., 57–64.

Hornung, M. (1985) Acidification of soils by trees and forests. *Soil Use and Management*, **1**, 24–8.

Howells, G. D. and Kallend, A. S. (1984) Acid rain – the CEGB view. *Chemistry in Britain*, May, 407–12.

Hutchinson, T. C. and Whitby, L. M. (1977) Effects of acid rainfall and heavy metal particulates on a boreal forest ecosystem near Sudbury smelting region of Canada. *Water, Air and Soil Pollution*, **7**, 421–8.

Huttunen, S. (1985) Ecophysiological effects of air pollution on conifers. In *Symposium on the Effects of Air Pollution on Forest and Water Ecosystems, Helsinki, April 23–24, 1985*, Foundation for Research of Natural Resources in Finland, Helsinki, 25–9.

Irving, P. M. (1983) Acid precipitation effects on crops: a review and analysis of research. *Journal of Environmental Quality*, **12**, 442–53.

Jarvis, N. J., Mullins, C. E. and MacLeod, D. A. (1983) The prediction of evapotranspiration and growth of Sitka spruce from meteorological records. *Annales Geophysicae*, **1**, 335–44.

Johannessen, M. and Henriksen, A. (1978) Chemistry of snow meltwater: Changes in concentration during melting. *Water Resources Research*, **14**, 615–19.

Johnson, D. W. and Henderson, G. S. (1979) Sulfate adsorption and sulfur fractions in a highly weathered soil under a mixed deciduous forest. *Soil Science*, **128**, 34–40.

Johnson, D. W. and Todd, D. E. (1984) Effects of acid irrigation on carbon dioxide evolution, extractable nitrogen, phosphorus and aluminium in a deciduous forest soil. *Soil Science Society of America Journal*, **48**, 664–6.

Johnson, N. M., Likens, G. E., Feller, M. C. and Driscoll, C. T. (1984) Acid rain and soil chemistry. *Science*, **225**, 1424–5.

Joranger, E., Schang, J. and Semb, A. (1980) Deposition of air pollutants in Norway. In *Ecological Impact of Acid Precipitation*, Drabløs, D. and Tollan, A., eds, SNSF Project, Oslo, 120–1.

Killham, K. and Firestone, M. K. (1982) Evaluation of accelerated $H^+$ applications in predicting soil chemical and microbial changes due to acid rain. *Communications in Soil Science and Plant Analysis*, **13**, 995–1001.

Killham, K. and Wainwright, M. (1984) Chemical and microbiological changes in soil following exposure to heavy atmospheric pollution. *Environmental Pollution (A)*, **33**, 121–31.

Killham, K., Firestone, M. K. and McColl, J. G. (1983) Acid rain and soil microbial activity: effects and their mechanisms. *Journal of Environmental Quality*, **12**, 133–7.

King, J. and Nicholson, I. A. (1964) Grasslands of the forest and subalpine zones. In *The Vegetation of Scotland*, Burnett, J. H., ed., Oliver and Boyd, Edinburgh, 168–231 (chapter 6).

Krug, E. C. and Frink, C. R. (1983) Acid rain on acid soil: a new perspective. *Science*, **221**, 520–5.

Langkramer, O and Lettl, A. (1982) Influence of industrial atmospheric pollution on soil biotic component of Norway spruce strands. *Zentralblatt für Mikrobiologie*, **127**, 180–96.

Lee, J. E. (1985) Effect of simulated acid rain on the chemistry of a sulfate-adsorbing forest soil. *Water, Air and Soil Pollution*, **25**, 185–93.

Likens, G. E., Bormann, F. H. and Johnson, N. M. (1972) Acid rain. *Environment*, **14**, 33–40.

Likens, G. E., Bormann, F. H., Pierce, R. S., Eaton, J. S. and Johnson, N. M. (1977) *Biogeochemistry of a Forested Ecosystem*. Springer-Verlag, New York.

Matzner, E. (1983) Balances of element fluxes within different ecosystems impacted by acid rain. In *Effects of Accumulation of Air Pollutants in Forest Ecosystems*, Ulrich, B. and Pankrath, J., eds, D. Reidel Publishing Co., 147–55.

Matzner, E. (1985) Effects of acid precipitation on soils – principles demonstrated in two forest ecosystems of North Germany. In *Symposium on the Effects of Air Pollution on Forest and Water Ecosystems, Helsinki, April 23–24 1985*, Foundation for Research of Natural Resources in Finland, Helsinki, 47–56.

Matzner, E. and Ulrich, B. (1981) Bilanzierung jährlicher Elementflusse in Waldöko-system im Soling. *Zeitschrift für Pflanzenernährung, Dungung und Boden-kunde*, **144**, 660–81.

McCaig, M. (1983) Contributions to storm quickflow in a small headwater catchment – the role of natural pipes and soil macropores. *Earth Surface Processes and Landforms*, **8**, 239–52.

McLaughlin, D. L. (1985) Air pollution, acid rain and forestry – a Canadian perspective. In *Symposium on the Effects of Air Pollution on Forest and Water Ecosystems, Helsinki, April 23–24 1985*, Foundation for Research of Natural Resources in Finland, Helsinki, 115–26.

McVean, D. N. (1964) Regional pattern of the vegetation. In *The Vegetation of Scotland*, Burnett, J. H., ed., Oliver and Boyd, Edinburgh, 568–78 (chapter 17).

Mecklenburg, R. A., Tukey, Jr H. B. and Morgan, J. V. (1966) A mechanism for the leaching of calcium from foliage. *Plant Physiology*, **41**, 610–13.

Miles, J. (1978) The influence of trees on soil properties. *Annual Report of the Institute of Terrestrial Ecology*, 1977, 7–11.

Miles, J. (1981) *The Effect of Birch on Moorlands*. Institute of Terrestrial Ecology, Cambridge.

Miller, H. G. (1983) Studies of proton flux in forests and heaths of Scotland. In *Effects of Accumulation of Air Pollutants in Forest Ecosystems*, Ulrich, B. and Pankrath, J., eds, D. Reidel Publishing Co., 183–93.

Miller, H. G. (1984a) Water in forests. *Scottish Forestry*, 165–81.

Miller, H. G. (1984b) Deposition–plant–soil interactions. *Philosophical Transactions of the Royal Society of London*, **B305**, 339–52.

Miller, H. G. (1985a) The possible role of forests in streamwater acidification. *Soil Use and Management*, **1**, 28–9.

Miller, H. G. (1985b) Acid flux and the influence of vegetation. In *Symposium on the Effects of Air Pollution on Forest and Water Ecosystems, Helskini, April 23–24 1985*, Foundation for Research of Natural Resources in Finland, Helsinki, 37–46.

Miller, H. G. and Miller, J. D. (1980) Collection and retention of atmospheric pollutants by vegetation. In *Proceedings of the International Conference on the Ecological Impact of Acid Precipitation*, SNSF Project, Norway, 33–40.

Morrison, I. K. (1983) Composition of percolate from reconstructed profiles of two Jack pine forest soils as influenced by acid input. In *Effects of Accumulation of Air Pollutants in Forest Ecosystems*, Ulrich, B. and Pankrath, J., eds, D. Reidel Publishing Co., 195–206.

Nilsson, S. I. (1983) Effects of soil chemistry as a consequence of proton input. In *Effects of Accumulation of Air Pollutants in Forest Ecosystems*, Ulrich, B. and Pankrath, J., eds, D. Reidel Publishing Co., 105–11.

Nilsson, S. I., Miller, H. G. and Miller, J. D. (1982) Forest growth as a possible cause of soil and water acidification: an examination of the concepts. *Oikos*, **39**, 40–9.

Overrein, L. N. (1972) Sulphur pollution patterns observed; leaching of calcium in forest soil determined. *Ambio*, **1**, 145–7.

Ovington, J. D. (1953) Studies on the development of woodland conditions under different trees. I. Soils pH. *Journal of Ecology*, **41**, 13–34.

Paces, T. (1982) Natural and anthropogenic flux of major elements from Central Europe. *Ambio*, **11**, 206–8.

Paces, T. (1985) Sources of acidification in Central Europe estimated from elemental budgets in small basins. *Nature* (London), **315**, 31–6.

Reid, J. M., MacLeod, D. A. and Cresser, M. S. (1981) Factors affecting the chemistry of precipitation and river water in an upland catchment. *Journal of Hydrology*, **50**, 129–45.

Ritchie, J. (1920) *The Influence of Man on Animal Life in Scotland*, Cambridge University Press.

Rodhe, H., Crutzen, P. and Vanderpol, A. (1981) Formation of sulphuric and nitric acid in the atmosphere during long-range transport. *Tellus*, **33**, 132–41.

Ronneau, C. and Snappe-Jacob, N. (1978) Atmospheric transport and transformation rate of sulphur dioxide. *Atmospheric Environment*, **12**, 1517–21.

Ronneau, C., Navarre, J. L., Priest, P. and Cara, J. (1978) A three-year study of air pollution episodes in a semi-rural area. *Atmospheric Environment*, **12**, 877–81.

Rosenqvist, I. Th. (1977) *Sur jord – Surt Vann* (Acid Soil – Acid Water). Ingeniørforlaget A/S, Oslo.

Rosenqvist, I. Th. (1978) Alternative sources for acidification of river water in Norway. *Science of the Total Environment*, **10**, 39–49.

Rosenqvist, I. Th. (1981) The importance of acid precipitation and acid soil in freshwater lake chemistry. *Vann*, No. 4, 402–9.

Rowell, D. L. and Wild, A. (1985) Causes of soil acidification. *Soil Use and Management*, **1**, 32–3.

Sasanow, S. (1985) Acid rain: a review of current controversy. *Soil Use and Management*, **1**, 34–7.

Scherbatskoy, T. and Klein, R. M. (1983) Response of spruce and birch foliage to leaching by acid mists. *Journal of Environmental Quality*, **12**, 189–95.

Skeffington, R. A. (1983) Soil properties under three species of tree in southern England in relation to acid deposition of throughfall. In *Effects of*

*Accumulation of Air Pollutants in Forest Ecosystems*, Ulrich, B. and Pankrath, H., eds, D. Reidel Publishing Co., 219–31.

Skiba, U. and Cresser, M. S. (1986) Effects of precipitation acidity on the chemistry and microbiology of Sitka spruce litter leachate. *Environmental Pollution (A)*, **42**, 65–78.

Skiba, U., Peirson-Smith, T. J. and Cresser, M. S. (1986) Effects of simulated precipitation acidified with sulphuric acid and/or nitric acid on the throughfall chemistry of Sitka spruce (*Picea stichensis*) and heather (*Calluna vulgaris*). *Environmental Pollution (B)*, **11**, 255–70.

Smith, R. T. and Taylor, J. A. (1969) The post-glacial development of vegetation and soils in northern Cardiganshire. *Transactions of the Institute of British Geographers*, **48**, 75–97.

Sorensen, N. A. (1984) Comment upon 'modelling streamwater chemistry' by Christopherson, N., Rustad, S. and Seip, H. M. *Philosophical Transactions of the Royal Society of London*, **B305**, 439.

Starr, M. R. (1985) Prescribed burning and soil acidification. In *Symposium on the Effects of Air Pollution on Forest and Water Ecosystems, Helskini, April 23–24 1985*, Foundation for Research of Natural Resources in Finland, Helsinki, 101–6.

Tabatabai, M. A. (1985) Effect of acid rain on soils. *CRC Critical Reviews in Environmental Control*, **15**, 65–110.

Taylor, J. A. (1985) Bracken encroachment rates in Britain. *Soil Use and Management*, **1**, 53–6.

Thompson, T. R. E. and Loveland, P. J. (1985) The acidity of Welsh soils. *Soil Use and Management*, **1**, 21–4.

Tilsley, J. E. (1977) Placosols: another problem in exploration geochemistry. *Journal of Geochemical Exploration*, **7**, 21–30.

Tomlinson, G. H. (1983a) Air pollutants and forest decline. *Environmental Science and Technology*, **17**, 246A–56A.

Tomlinson, G. H. (1983b) Dieback of red spruce, acid deposition, and changes in soil nutrient status, a review. In *Effects of Accumulation of Air Pollutants in Forest Ecosystems*, Ulrich, B. and Pankrath, J., eds, D. Reidel Publishing Co., 331–41.

Tukey, H. B. (1970) The leaching of substances from plants. *Annual Reviews of Plant Physiology*, **21**, 305–29.

Tukey, Jr, H. B. (1980) Some effects of rain and mist on plants, with implications for acid precipitation. In *Effects of Acid Precipitation on Terrestrial Ecosystems*, Hutchinson, T. C. and Havas, M., eds, Plenum Press, New York, 141–50.

Tyler, G. (1983) Does acidification increase metal availability and thereby inhibit decomposition and mineralisation processes in forest soils? In *Ecological Effects of Acid Deposition*, National Swedish Environment Protection Board, Report PM1636, 245–56.

Ulrich, B. (1983a) A concept of forest ecosystem stability and of acid deposition as driving force for destabilization. In *Effects of Accumulation of Air Pollutants in Forest Ecosystems*, Ulrich, B. and Pankrath, J., eds, D. Reidel Publishing Co., 1–29.

Ulrich, B. (1983b) Soil acidity and its relations to acid deposition. In *Effects of Accumulation of Air Pollutants in Forest Ecosystems*, Ulrich, B. and Pankrath, J., eds, D. Reidel Publishing Co., 127–46.

Ulrich, B. (1983c) Interaction of forest canopies with atmospheric constituents: $SO_2$, alkali and earth alkali cations and chloride. In *Effects of Accumulation of Air Pollutants in Forest Ecosystems*, Ulrich, B. and Pankrath, J., eds, D. Reidel Publishing Co., 33–45.

Underwood, J. K., Vaughan, H. H., Ogden III, J. G. and Kerekes, J. J. (1983) Acidification of Nova Scotia lakes. III: Patterns of $SO_4$ and $NO_3$ deposition, and $SO_4$ concentrations in rural and urban lakes. Distributed at a poster presented at the September 1983 meeting of the Royal Society of London on 'Ecological Effects of Deposited Sulphur and Nitrogen Compounds'.

van Breemen, N., Burrough, P. A., Velthorst, E. J., van Dobben, H. F., de Wet, T., Ridder, T. B. and Reijnders, H. F. R. (1982) Soil acidification from atmospheric ammonium sulphate in forest canopy throughfall. *Nature* (London), **299**, 548–50.

van Breemen, N. and Jordens, E. R. (1983) Effects of atmospheric ammonium sulphate on calcareous and non-calcareous soils of woodlands in the Netherlands. In *Effects of Accumulation of Air Pollutants in Forest Ecosystems*, Ulrich, B. and Pankrath, J., eds, D. Reidel Publishing Co., 171–82.

van Breemen, N., Mulder, J. and Driscoll, C. T. (1983) Acidification and aklalinization of soils. *Plant and Soil*, **75**, 283–308.

van Breemen, N., Driscoll, C. T. and Mulder, J. (1984) Acidic deposition and internal proton sources in acidification of soils and waters. *Nature* (London), **307**, 599–604.

Vaughan, H. H., Underwood, J. K., Kerkes, J. J. and Ogden III, J. G. (1983) Acidification of Nova Scotia lakes. IV. Response of diatom assemblages in rural lakes. Distributed at a poster presentation at the September 1983 meeting of the Royal Society of London on 'Ecological Effects of Deposited Sulphur and Nitrogen Compounds'.

Watt Committee on Energy (1984) *Acid Rain, Report No. 14*, The Watt Committee on Energy Ltd.

Wiklander, L. (1975) The role of neutral salts in the ion exchange between acid precipitation and soil. *Geoderma*, **14**, 93–105.

Wood, T. and Bormann, F. H. (1975) Increases in foliar leaching caused by acidification of an artificial mist. *Ambio*, **4**, 169–75.

# 4

○ ○ ○ ○ ○ ○ ○ ○ ○ ○ ○ ○ ○ ○ ○ ○ ○ ○ ○ ○

## Experimental methods in acidification research

The success of the outcome of any experimental study, whether in the field or in the laboratory, depends upon the input of thought and effort at the experimental design stage. Particularly for field-based studies, lasting perhaps for a year or more, it is imperative to make sure that the most appropriate specific question is being asked, and further that the data being obtained will provide the answer required. An adequate knowledge of the variability of measured parameters in space and time is therefore critical when deciding upon sampling strategy. Hopefully the background information required at the planning stage of water acidification investigations may be found in Chapters 2 and 3.

For a field-based study, six major considerations must be addressed:
- (i) Choice of a suitable site.
- (ii) Sample type and sampling frequency and spacial density.
- (iii) Duration of the study.
- (iv) Financial constraints.
- (v) Analytical facilities available.
- (vi Relationship to other investigations.

Obviously to some extent these factors are self-explanatory and inter-related. Some of them include a host of minor considerations. Site choice, for example, obviously depends upon the purpose of the investigation, but may also be influenced by site area (in relation to (ii), (iv) and (v) above), accessibility, particularly during winter months in remote areas, the need for electrical power, permission to install necessary monitoring and/or sampling structures, the risk of vandalism, etc. Often not every demand is fully satisfied, and a degree of compromise may be unavoidable. A flexible approach is often advisable, particularly in the case of sampling frequency and extent of replication. Initially therefore it may be necessary to collect samples more frequently and with greater replication than ultimately required, simply to establish the patterns of temporal and spacial variability to ensure that the sampling strategy eventually adopted is adequate. The purpose of this section is primarily to highlight the major points from Chapters 2 and 3 that must be borne in mind at the experimental design stage.

## Possible sampling objectives

It is very important to define very precisely *why* samples are being collected. All too often in acid rain research, project aims have been couched in rather vague terms. Sometimes, but not always, this has been done deliberately, to leave scope for flexibility in a project. Occasionally, however, vague terms reflect woolly thinking or lack of detailed planning. The sort of problems that might arise become clearer when a specific example is considered.

'A study will be made of the susceptibility of surface waters in the area of acidification'. What does this mean? What will be studied, rivers or lakes? Will the distribution of pH in lakes or along streams be studied? What is most interesting and relevant, river baseflow acidity or storm flow acidity? Is the discharge (quantity of water) important as well as water pH? Is rate of change in pH throughout storms important? Do we need to consider the 'average' storm, or the worst storm likely each year, each ten years or each hundred years? The answers to these questions depend on why the study was proposed, and the response to yet more questions. What is really important, the fundamental mechanisms involved, effects on fish, effects on water pipes, likely effects over the next decade, possible effects over the next century, etc.? These questions must be answered first, otherwise there is a risk of wasting research funding and obtaining an inadequate set of data. It should be further realised that so far here only water in rivers and lakes has been considered, not the impact of vegetation, soil, geology, topography, etc. These factors, too, may need to be taken into account. Sanders and Adrian (1978) suggested, quite correctly, that in many cases professional judgement and cost constraints provide the basis for choosing a particular sampling frequency.

## Sampling frequency

The dynamic nature of the chemical, physical and biological processes occurring at the earth's surface mean that environmental scientists are, more often than not, sampling a continuously changing system. The changes may be seasonal or diurnal, or on a longer or shorter time scale. As mentioned earlier, dramatic changes in precipitation chemistry may occur over a few minutes during a single shower, or in river-water chemistry in an upland drainage basin over less than an hour or a few hours (depending upon catchment size). Accurate assessment of $H^+$ outputs in an upland stream may therefore require hourly or even more frequent sampling over the duration of storms and a few hours following cessation of precipitation events. Rapid snowmelt, too, may require frequent sampling. Because of the nature of flow duration curves in upland catchments in moderate to high precipitation areas (see Chapter 2), even weekly sampling may capture only one or two storm events, and give no indication of a *potential* low pH discharge (Edwards *et al.*, 1984). Klein (1981), for example, noted the importance of a high sampling frequency for river water during storms in a study of the flushing effect for a small Yorkshire catchment. Spot sampling on one or two random single days must therefore be regarded as being almost valueless for any acidification-related purpose. Worst of all perhaps is the spot sampling exercise completed on a 'lovely sunny day – ideal for field work!'. Such a naive approach is rare, but not totally unheard of. Spot sampling during an abnormally prolonged wet period may however have some value, as discussed in Chapter 2. River-water chemistry under such conditions reflects low soil base status and hydrological pathways.

On the other hand, for annual budgets to be used to estimate geochemical weathering rates, weekly sampling and chemical analyses of both precipitation and river water, coupled to continuous discharge monitoring, may be perfectly adequate (Edwards *et al.*, 1984). Estimates of annual mean solute concentration do tend however to show increased error as the sampling interval is increased (Rainwater and Avrett, 1962). Changes in river-water chemistry with time tend to be greater in smaller catchments, all other things being equal. Therefore the number of samples required to give comparable reliability for subsequent statistical analysis also tends to increase with decreasing catchment area (Sanders and Adrian, 1978).

Whereas numerous studies have included approximately hourly sampling of rivers, and such work has given appreciable insight into the hydrology of upland catchments, relatively few studies have examined precipitation chemistry at high sampling frequency. Acidity may however change substantially during a single storm (see e.g. Edwards *et al.*, 1984). In certain types of steep, rocky terrain, such changes could be significant in the present context, and they probably warrant more attention than they have thus far received. It would be inappropriate to discuss sampling frequency further here, but the interested reader should consult papers by Frere (1971), Montgomery and Hart (1974) or Edwards *et al.* (1984).

## Sampling river and lake water

Numerous options are available for river and lake-water sampling. The simplest approach is to take samples at specified time intervals (e.g. daily, weekly, monthly, etc.), irrespective of discharge. Alternatively, sample collection may be automatically linked to either precipitation or, more usually for streams, to discharge. Indeed, for following changes in river-water chemistry during precipitation or thaw events, such an approach may be essential. Commercially available automatic samplers may collect samples at specified times or collect sub-samples at shorter time intervals and combine them to yield composite samples. For example, sets of four 15-minute samples may be auto-matically combined to give combined hourly samples. Sampling is not necessarily most appropriately done at pre-selected time intervals. Thus it may be preferable to link the sampling interval automatically to dis-charge, with relatively much shorter intervals at higher discharge when water chemistry is most likely to be changing rapidly. Claridge (1975) found a system that operated at a frequency proportional to discharge to be very satisfactory.

Assessment of output yields in rivers requires a knowledge of two parameters, concentration and discharge. The latter requires a form of gauging device such as a flume or weir, or alternatively dilution gauging may be used for flow assessment. Discussion of relevant techniques may be found in hydrology textbooks. For small streams, flumes are rela-tively inexpensive, but cost rises rapidly with increasing catchment size and discharge range to be covered. Stream sediment load must be borne in mind in some investigations, if total elemental outputs from a site are to be assessed. Even if a measure of sediment loss is not required, steps may be necessary to minimise the risk of silting up of the discharge monitoring system (Newson, 1981). Automated stream sediment sampling also may present problems (Nordin and Dempster, 1963; Schneider and Angino, 1980).

If autosamplers are available, some thought must be given as to how they may be put to best use. We have made extensive use of a system similar to that described by Walling and Teed (1971), which samples at pre-set time intervals once triggered by rising water level in a stilling well. A major disadvantage of this system is the lack of a reference baseflow stream sample, which we overcome to some extent by collecting weekly spot samples. The use of a tipping-bucket rain gauge to provide an electric pulse which initiates and controls the action of an autosampler has been suggested by Martin and White (1982), the sampler being triggered at a pre-determined rain intensity. Stream response times to individual storms for a particular catchment vary seasonally according to soil moisture status. Both the systems described above may miss long-duration low-intensity storm events. The latter system also fails to sample during periods of snowmelt, an important consideration for many of the catchments relevant to this book.

In our experience, automatic sampling during really severe winter conditions is fraught with difficulties. Risk of battery failure may be minimised by providing some form of heating (frost protection) for the autosampler itself. However it is much more difficult to overcome the problem of ice formation at the sample inlet point in the stream. Safe sampling of water beneath frozen lake surfaces may also present problems. As discussed elsewhere, thermal stratification may be pronounced in frozen lakes and appreciable chemical gradients may occur. In rivers, mobilised chunks of ice may cause damage to discharge measurement and/or sampling systems. Unfortunately these problems tend to occur at the very times when accessibility to remote catchments for manual sampling may be severely restricted.

## Characterisation of wet and dry deposition inputs

The origins of acidifying pollutant deposition have already been discussed in Chapters 2 and 3. It will be recalled that wet deposition involves the transfer of elements from the atmosphere in aqueous solution or suspension (rain, sleet, snow, etc.). Dry deposition involves the direct transfer of gaseous and particulate material from the atmosphere to surfaces (excluding wet deposition). Dry deposition is a continuous process, which means that both types of input may occur simultaneously. This leads to problems in the field in distinguishing between contributions from the two sources.

*Monitoring of dry deposition inputs*

Accurate quantification of dry deposition inputs of any species is extremely difficult to achieve. Many factors interact to complicate direct measurement. Large differences in catching efficiencies are exhibited by differing surfaces and surface geometries, and hence by different vegetation types. Thus heather moorland may be expected to 'catch' much less dry deposited material than a forested site. It follows that seasonal differences in catching efficiencies are to be expected, especially when comparing coniferous and deciduous sites, and even when comparing trees of differing age. Moreover, a sizeable portion of the potential input is water soluble, and therefore higher deposition levels can be expected when soil and vegetation surfaces are moist (Fowler and Unsworth, 1974). A further complication which may arise when attempts are made to extrapolate from one site to another is the possibility of altitudinal effects. For example, Georgii and Lenhard (1978) found that the ratio of aerosol $NH_4^+$ to $NH_3$ gas diminished rapidly with height, while they reported a more uniform distribution for $SO_4^{2-}$. Even if it was possible, therefore, to erect a chemically and biologically inert structure with an appropriate collecting efficiency under one set of conditions, the surface thus created might be wholly inappropriate under other circumstances. Researchers in this area are thus beset with difficulties.

In spite of this, filter gauges with a capacity for enhanced aerosol impaction have been used. Nihlgård (1970) noted increased element concentrations beneath a framework of plastic netting compared in an open-sited gauge. Miller and Miller (1980) used cylinders of polyethylene-coated wire mesh above rain collectors, and compared concentrations to those for the more aerodynamic Nipher shielded gauges (which are designed to minimise aerosol impaction) at a number of forested sites. Such an approach is valuable for qualitative or even semi-quantitative confirmation of trends in dry deposition, for example with altitude or with distance from an input source, etc. However, little progress has apparently been made with their calibration, and it is difficult to see how they can ever be used to provide reliable quantitative estimates of dry deposition fluxes. Even if catchment sodium and chloride balances suggest that they may be useful for estimation of maritime-derived aerosol inputs, it does not follow that they are equally appropriate for dry deposition inputs of other origins.

Some direct information is available on the typical ranges and ratios of elements present in particulates of specified sizes (Kadowaki, 1976; Brosset, 1978). Such studies have highlighted the importance of particle size in the context of acidifying pollutant deposition. This type of characterisation of particulate dry deposition inputs is achieved using cascade impactors, which fractionate particulates from a known volume of air into size categories. Harrison and Pio (1983a, b) have described such a system. Several versions are commercially available. Harrison and Pio (1983c) also described a system for trapping of gaseous pollutants with various impregnated filters after pre-extraction of atmospheric aerosols using Teflon filters. If analytical facilities are available, direct determination of $SO_2$, $H_2S$ and $NO_3$ in the field presents no great difficulty. Many estimations of particulate and gaseous fluxes to canopies have been based upon deposition velocity values derived from laboratory experiments (Chamberlain, 1975). Unsworth (1980) has reviewed progress in our understanding of dry deposition of gases and particles on to vegetation, and quoted typical deposition velocities. Reported values cover a considerable range.

Indirect methods may also be used to quantify inputs of some elements from dry deposition using long-term catchment budgets (Mayer and Ulrich, 1974; Reid *et al.*, 1981). Even these are not without problems; chloride balance may possibly be used to estimate sodium and magnesium in areas with maritime domination, but sulphate balance is of limited value because of sulphate adsorption and $H_2S$ retention in anaerobic peats. Thus indirect methods are of limited applicability in the present context.

Filter-based methods suffer from two additional problems. The first is that, if the filter material is inert, the neutralisation, ion exchange and biological reactions which might occur on impaction with a biologically active plant surface are eliminated. They may therefore seriously misrepresent the actual acid fluxes. The second is that they may collect blown snow, which complicates interpretation of results obtained when blown snow is significant. It appears therefore that calculated estimates based upon deposition velocities may be most reliable, if not as accurate as desirable.

*Monitoring of wet deposition inputs*
Although more familiar, wet deposition collection is also not without its problems, despite the long history of rainfall monitoring (Ward 1975). Significant sources of error (especially at exposed sites typical of the present discussion) can result from wind deflection, and hence turbulence and eddying, produced by the presence of the gauge and surrounding vegetation. Increased air flow results in rain being carried past rather than falling into the gauge. The value of ground-level recorders in certain situations has been demonstrated (Strangeways, 1982). Lack of a widely accepted standard rain gauge makes valid inter-site comparisons of precipitation data difficult.

Precipitation samples tend to be collected over a pre-determined time period. As suggested by Whitehead and Feth (1964), a continuously open rain gauge will collect a varying proportion of material from dry depositional sources. They therefore applied the term 'bulk precipitation' to samples collected in this manner. Cape *et al.* (1984) suggested the dry deposition of material, e.g. $SO_2$, on to precipitation collectors is non-linear with time and may account for 15–35% of the measured total sulphate. While less of a problem with catchment budget accounts, this 'contamination' of wet deposition samples is more of a problem when trying to resolve the two input components. With this in mind 'wet only' precipitation collectors are available. An automatically removable lid uncovers the protected gauge with the onset of precipitation (Galloway and Likens, 1978). If, however, such gauges are excessively bulky, their effect on air flow may itself cause errors.

As with river water there are several considerations relevant to the collection of precipitation samples. Lewis and Grant (1978) suggested that the method of precipitation collection and processing greatly affects the apparent relative contributions of the three identifiable fractions: dissolved materials in aqueous precipitation, the water-soluble component of dry deposition, and the water-insoluble component of either wet or dry precipitation. Filtering may obviously be important, and acidifying a sample for storage may well dissolve particulate material. They also noted the importance of an adequate rain gauge size in order to obtain sufficient samples for all the analytical determinations required.

The number of rain gauges required to adequately predict inputs to a given area should also be considered. Richter *et al.* (1983) suggested that a large number of collectors were required to make even moderately precise estimates within the 500 ha Santee watershed in the eastern United States. Reynolds (1984), studying the upper Wye in mid-Wales, UK, reported little effect of the aspect and elevation of collectors on solute concentrations and suggested one sampler would be adequate (10% error). He did note, however, that differences in the actual rainfall catch within the watershed were the major cause of variation in solute load. There is a tendency for a negative correlation between rainfall amount and solute concentration (Cryer 1976), which suggests that volume-weighted solute concentrations should be used for comparison of inputs. The change in precipitation chemistry with time during individual storm events has been the subject of a number of papers. Two basic designs of fractionating sampler have been employed, one based upon fixed volume collection, the other based upon pre-set time intervals regardless of rainfall intensity. Numerous examples of both types may be found in the references we have listed elsewhere (Edwards and Cresser, 1985). Fixed-volume samplers always provide sufficient sample for analysis, but they have a limited capacity and must be used in conjunction with a tipping-bucket gauge to allow for timing of the samples. Fixed time intervals circumvent the latter problem, but may provide insufficient sample for analysis.

Precipitation sampling in severe winter weather may pose two principal problems. The sample may freeze in the gauge, making removal and immediate measurement difficult or even impossible. We avoid this problem by using double polythene bags, the upper one with a slit bottom, to line suitable drums. The lower bag is sealed and brought to the laboratory, for thawing, volume measurement, and chemical analysis, and replaced by a new bag. Drifting snow presents an additional problem, to which there is no simple answer. This complication has been mentioned already at several points in the text.

*Throughfall and stemflow*
Throughfall may be measured using conventional rain gauges, or, to provide more integrated samples, with 2-m lengths of guttering suspended above the ground and sloping towards a collector. Sets of three such gutters may yield an excellent representative sample of throughfall water. However, their collection efficiency should be checked against that of more widely accepted designs if valid comparisons of results with those of other studies are to be made. They may be susceptible to evaporation loss.

A complication with throughfall is the fate of plant litter deposited in the collector. Often this is removed with an in-line filter, the filter fibre or paper being changed as each sample is collected. The filtration rate must be capable of coping with the maximum water input rate. If precipitation samples are collected at infrequent intervals, some decomposition of the trapped litter is unavoidable, and it may slightly modify the solute chemistry of subsequent throughfall drainage through it. It is often argued that this is of little consequence, because such a process occurs quite naturally at the soil surface anyway. Whether this is a valid argument or not depends on the purpose for which samples are being collected. Debris may be a particular problem under smaller plants such as *Calluna vulgaris* or sphagnum. Filtration is then essential. A system used by us and our colleagues for small plants is shown in Fig. 4.1. Water is removed by syringe and triplicate samples combined to provide a more representative sample.

Stemflow is usually collected by attaching a spiral gutter to a tree trunk. Various gutter materials have been used, from silicon rubber tubing bisected lengthwise, to purpose-moulded flexible vinyl compounds (e.g. Miller and Miller, 1976), to waterproof-coated expanded polyurethane on a wire-mesh support. Mastic is usually used between the gutter and the tree to allow for expansion. Careful consideration must be given to the selection of trees if the data is to be extrapolated to give ha$^{-1}$ inputs and outputs.

Further discussion of the approaches adopted by different workers to studying wet deposition is beyond the scope of this volume. The interested reader should consult the report of the proceedings of a workshop on methods for studying acid rain precipitation in forest ecosystems, held at the Institute of Terrestrial Ecology in Edinburgh in 1977 (Nicholson *et al.*, 1980).

Fig. 4.1. System for collection of throughfall under small plants such as *Calluna vulgaris* or sphagnum.

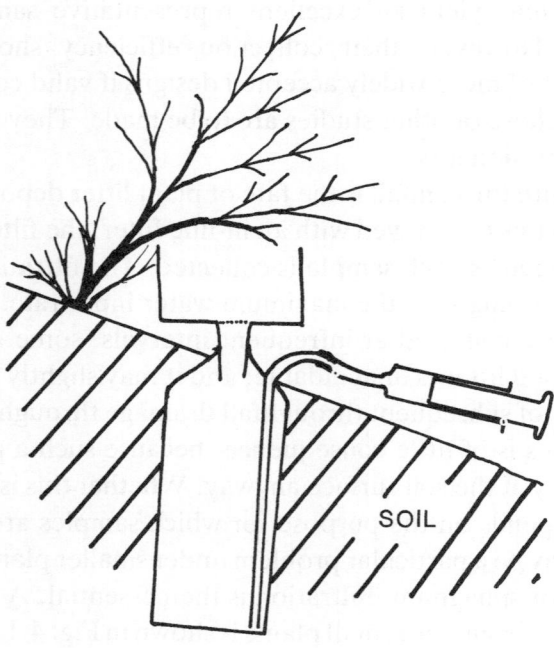

## Lysimeters in 'acid rain' research

As discussed in Chapter 2, water solute chemistry may be substantially and rapidly modified as a consequence of passage through a
soil matrix. Soil water composition is therefore often of as much interest
as precipitation or river or lake-water chemistry. Numerous techniques
are available, but they fall into two broad categories. The first group
includes methods in which water is allowed to drain freely from a soil,
in which case the soil, at least at the sampling point, must be saturated.
The second set includes techniques in which water is removed by application of increased or reduced pressure to overcome the tensional forces
holding the water in the soil matrix. The latter group includes destruc-
tive sampling techniques whereby soil samples are brought back to the
laboratory for water removal. These are less popular, for obvious
reasons, than methodology which allows water samples to be collected
at regular intervals with minimal disturbance once the sampling device
has been installed. Such sampling systems are collectively known as
'lysimeters'.

*Tension lysimeters*

As the name suggests, tension lysimeters rely upon applied suction to
overcome soil water tension. There are various types of sampler available, but all fall into one or other of two categories, either ceramic plates
inserted below the soil horizons of interest or ceramic cups placed into
holes obtained by augering to the required depth. Exponents of these
devices argue that they are potentially useful, and once in place their
influence is minimal in terms of further disturbance to the soil/plant/
water system. They might therefore seem ideal for monitoring changes
in solute chemistry with time. They are not without their problems however. Three aspects in particular must be considered, namely the introduction of contaminants, modifications of solute chemistry as a result of
adsorption, and temporal variability in the sampling zone resulting
from changes in soil moisture status.

Bringmark (1980), in a study of iron movement through podzol profiles, used a pumping system to maintain constant suction, and recommended a hydrochloric acid wash procedure, followed by careful
dionised water rinsing prior to use. However, Wolff (1967) reported the
release of calcium, sodium, magnesium, bicarbonate and silicate after
cleaning with dilute hydrochloric acid. It is clearly necessary to check
carefully for contamination problems in the laboratory prior to field

use. This applies not just to the cups or plates themselves, but also to the other materials, especially sealants, used in sampling system construction.

Mayer (1971) noted that, while most of the cations of interest passed unchanged through this system, retention of sulphate and phosphate by aluminium leached from the ceramic material occurred. It is generally tacitly assumed that, once inside the ceramic cup or through a ceramic plate, water chemistry remains unchanged. This may not be the case in practice. Outgassing of carbon dioxide, for example, may lead to a pH rise, precipitation of aluminium and iron oxyhydroxides, and anion and trace-element adsorption on the precipitate.

It is essential to have good hydraulic continuity between the cup and/or plate and the adjacent soil, and this may be difficult to attain and maintain. This, and the seasonal and other temporal changes that occur in soil moisture content, and hence hydraulic conductivity, cause variations in the volume and zone of soil sampled under constant low pressure. As a result the reliability of such systems for assessing field soil water and its solute composition has been questioned (Hansen and Harris, 1975; Haines *et al.*, 1982; Nagpal, 1982). Some workers are of the opinion that it is best to almost match soil tension and applied tension, but such a sampling strategy is more complex to operate.

*Zero-tension lysimeters*
Zero-tension lysimeters rely upon the collection of water seeping from a free face and therefore sample only saturated flow. Water from individual soil horizons on sloping sites may be monitored by digging a pit and inserting lengths of guttering or plastic sheeting upslope along the soil horizon interface. Flow through surface litter or organic horizons may also be monitored in this way. The requirement of saturated flow for zero-tension lysimeters to function is met because soil at the pit face becomes saturated, with a wedge of saturated soil extending upslope from the pit face into zones which might not have become saturated in the absence of the soil pit (Atkinson, 1978). The extent and thickness of the wedge are dependent upon the water flux from upslope, and change as the flux changes (Hewlett and Hibbert, 1963). Pits may also receive water from areas not directly upslope due to distortion of the hydraulic potential network. This effect may be minimised by sealing the free pit faces by replacing the soil once the gutters and collector tubes have been installed. Unless the slope is moderate to steep, this may involve removing the samples from the storage collector vessel when required

with a syringe. Weyman (1970, 1974) circumvented the problem by choosing a site where a saturated wedge existed already, i.e. a stream bank.

*Soil monoliths*

Intact soil monoliths of varying sizes have often been used in leaching studies and may be regarded as a particular type of zero-tension lysimeter. To sample water from different horizons involves collecting a series of monoliths of differing depths. Collection is not always straight-forward, particularly in very stony soils. We use large-diameter plastic drain-pipes with bevelled edges to facilitate insertion into the soil. The pipe is pushed slowly downwards, surrounding material being removed from around the pipe as it is pressed downwards. Contact with stones other than very small ones unfortunately means starting again else-where. Checks can be made by use of a steel wire probe just outside the proposed sampling area. Unfortunately this gives no indication of sub-stantial stones within the monolith, although these can of course be checked for after the experiment is complete. If replicate cores are to receive different treatments, it is best to give them all the same control treatment first, to make sure they behave in a similar fashion, before commencing a long-term diverse treatment programme. Once obtained, the lysimeters may be used in the field or removed carefully to the laboratory for simulation-based studies. In either case there is the great advantage of being able to apply several different treatments under otherwise similar conditions. Abrahamsen *et al.* (1977) for example supplemented 'natural' rain with artificial treatments to com-pare the resulting leachates.

We have used intact cores (monoliths) and also repacked soil cores, leaching them with rain of different pH and acidifying pollutant com-positions. The 'rain' was applied from a purpose-built pneumatic nebuliser system similar to that employed by Brunstad and Njøs (1980). It is best to control the precipitation rate either by use of a peristaltic pump or via gravity feed via a flow-restricting capillary, rather than to depend upon the nebuliser pumping action. The spray must not be too fine or much of the 'precipitation' will be lost by deflection. Using a fine spray is however more suitable for this type of study than keeping a con-stant head or using continual dripping, which may lead to structural problems. Nielsen and Biggar (1962) have shown that continuous pond-ing of water on the soil surface is inferior to rainwater for leaching of soil solutes. Water addition should be as close as possible to natural conditions.

Repacked columns are very convenient in so far as they greatly facilitate the preparation of replicates and a large number of cores for several diverse treatments. However, the dramatic modification to structure may cause problems, and it is important that the soils are not dried prior to use. Except in long-term studies, reconstituted cores eliminate any possibility of incorporating vegetation effects. They are very useful for obtaining an indication of soil properties, such as anion adsorption, which vary down the profile. Both repacked and intact core lysimeters suffer from the major drawback that they allow no possibility of lateral flow which, as has been seen in Chapters 2 and 3, may be a crucial factor in the regulation of river and lake-water chemistry.

## Other uses of simulated rain

Aside from the use of simulated rain in conjunction with lysimeters to study soil/water interactions, simulation techniques have also been extensively used to examine precipitation/vegetation interactions and the nature of the reactions involved (Skiba *et al.*, in press). The major merit of such techniques is that they allow elimination of all variables except the one under investigation, which may be for example pH, acid type, precipitation intensity, precipitation duration, etc.

## Sample contamination and storage

Care should be taken when designing sample collection systems to ensure that contamination and adsorption losses are negligible. This has already been mentioned briefly in relation to ceramic cups and plates for tension lysimeters, but other materials, too, may cause problems, and thorough washing of all components with deionised water is essential prior to routine use. We use inert plastic materials whenever possible. Water from natural sources is biologically active, and growth of micro-organisms may convert elements in a true solute to a suspension form. Filtration may then remove species initially present in the solution. Interconversion between nitrogen species may also occur, with an associated pH change. Growth of algae in transparent or translucent tubing and/or bottles may be a particular problem, but may be minimised by blackening the external surfaces. It is best to filter in-line, to prevent serious contamination from insects or litter during prolonged contact with samples. Gross contamination, from bird droppings for example, renders a sample useless.

Sample storage may require addition of a preservative such as chloroform or a mercury salt, provided these do not interfere in the subsequent analysis. Samples are best stored at low temperature, but should not be frozen. Freezing causes concentration of solute in the central water core, an effect which is often clearly visible in organic-rich river-water samples. It may thus lead to irreversible precipitation reactions. It is also advisable to keep sample storage times to a minimum whenever possible, and indeed continuous *in-situ* field analysis must be regarded as a highly desirable objective whenever the necessary facilities are available. 'On-line' pH determination is potentially very attractive in the present context, but is rarely used in practice at remote sites because of the many practical difficulties. As a second-best alternative, pH, bicarbonate and conductivity should be measured immediately samples are brought into the laboratory.

If storage is unavoidable, careful thought should be given, too, to the choice of bottles used and to washing procedures if contamination and adsorption losses are to be avoided. Mart (1979) and Patterson and Settle (1976) recommended a hot acid pre-wash for polyethylene bottles prior to 'normal' acid washing. It has been suggested by Nurnberg *et al.* (1976) that such a treatment could activate the bottle surface and give rise to trace-metal adsorption losses. We use a cold acid (1M HCl) treatment for new polyethylene bottles, followed by thorough deionised water rinsing, and storage prior to use full of distilled water. At sampling, the distilled water is emptied, and the bottle is carefully rinsed with sample water prior to filling. Trace-metal ions in solution tend to absorb strongly on to glass and to a lesser extent on to polyethylene (Florence, 1982). Glass-wool filters should be avoided for this reason, especially because of their high surface area. Carrick and Sutcliffe (1982) reported no change in the concentrations of sodium, potassium, calcium, magnesium or chloride in lake-water samples stored for eight months in polyethylene bottles at 15–22 °C. Losses tend to be significant mainly for the trace elements, and are particularly a problem in speciation studies. However occasionally we have in the past experienced microbial growth problems in river-water samples even for major elements. These are much less common for briefer, lower-temperature storage.

## Modelling and water acidity

It would be inappropriate in a text of this length to delve in depth into the possible input of modelling techniques in water acidification research. A complete book has been written on this aspect alone (Schnoor, 1984). Broadly speaking, four approaches have been adopted.

  (i)  Pollutant trajectory modelling.
  (ii)  Hydrological pathway modelling.
  (iii)  Modelling of rates of depletion of soil bases.
  (iv)  Fundamental process level modelling.

The first, atmospheric trajectory modelling, is particularly well advanced and useful. Trajectory modelling incorporates climatological data and the diverse chemical reactions and interactions which occur in the atmosphere. Well-developed systems allow reliable prediction of the impact of increases or decreases in the loadings of acidifying pollutants into the atmosphere in a specified region upon the wet and dry deposition loadings elsewhere. In the long term, such knowledge should be valuable, when used in conjunction with other models, for prediction of changes in soil and, ultimately, surface water acidity in response to changes in global patterns in fossil fuel combustion.

The second useful model group contains the hydrological models or sub-models, such as the Birkenes model (Christophersen and Wright, 1981; Christophersen *et al.*, 1982). The impetus in this and similar models has been the desire to explain the day-to-day variations in stream-water chemistry using relatively simple concepts of catchment hydrology. In the case of the Birkenes model, the catchment is treated as a two-reservoir system. Water fluxes between the reservoirs themselves and to and from the atmosphere and as drainage water are considered. Chemical processes, such as sulphate adsorption and desorption, may then be included to produce the final model for chemical behaviour. Establishment of such models requires a substantial input of both meteorological and chemical data in the first instance, but they may then be used to predict future behaviour often with impressive accuracy (Schnoor, 1984).

One limitation of such models is that the two-reservoir concept effectively encompassses the net result of a whole selection of complex interacting physical processes, as should be clear from Chapters 2 and 3. Thus their predictive capacity during periods of particularly abnormal weather may not be totally reliable. This could happen, for example, as a consequence of processes or hydrological pathways that are usually insignificant suddenly becoming important. A further limitation is that, because they are derived from concepts of net processes, rather than at the fundamental process level, it is not easy to reliably extrapolate from one catchment to another.

While models in the second category are useful for short-to-medium-term predictions of water quality, they must be used with care when making longer-term predictions unless the model includes relevant soil factors. These include changes in organic horizon pH and thickness, soil structural changes (possibly resulting in part from vegetational changes), and erosion effects. It is possible, however, to predict the rate of depletion of base cations from surface soils, either from geochemical weathering rate studies or from comparisons of parent material and present soil mineral chemical composition for mineral soil of known age (e.g. Edwards *et al.*, 1985). Such simple models allow quantification of long-term soil acidification rates to provide a background against which acid-deposition effects may be assessed.

The final approach to modelling involves examination at the fundamental process level throughout. Theoretically this is attractive because, provided all significant processes are considered, the model should ultimately become universal. In practice, however, the amount of input data required would be so high it would be virtually impossible to exploit the potential flexibility of the system. Therefore some simplifying 'black-box' assumptions will inevitably be required for a 'universal' model to be of genuine practical value. As yet, however, even the parameters that should be included in any such model are not widely agreed. We shall return to this problem in the final chapter. For further information the reader should consult the useful book edited by Schnoor (1984).

## References

Abrahamsen, G., Horntvedt, R. and Tvelh, B. (1977) Impacts of acid precipitation on coniferous forest ecosystems. *Water, Air and Soil Pollution*, **8**, 57–73.

Atkinson, T. C. (1978) Techniques for measuring subsurface flow on hillslopes. In *Hillslope Hydrology*, Kirkby, M. J., ed., Wiley, New York, 73–120.

Bringmark, L. (1980) Ion leaching through a podzol in a Scots pine stand. In *Structure and Function of Northern Coniferous Forests – An Ecosystem Study*, Persson, T., ed., *Ecological Bulletin (Stockholm)*, **32**, 341–61.

Brosset, C. (1978) Water-soluble sulphur compounds in aerosols. *Atmospheric Environment*, **12**, 25–38.

Brunstad, K. and Njøs, A. (1980) Simulation of flow patterns and ion-exchange in soil percolation experiments. Part 1. Tracer experiments and flow model. *Water, Air and Soil Pollution*, **13**, 459–72.

Cape, J. N., Fowler, D., Kinnaird, J. W., Paterson, I. S., Leith, I. D. and Nicholson,

I. A. (1984) Chemical composition of rainfall and wet deposition over northern Britain. *Atmospheric Environment*, **18**, 1921–32.

Carrick, T. R. and Sutcliffe, D. W. (1982) Concentrations of major ions in lakes and tarns of the English Lake District (1953–1978). *Freshwater Biological Association Occasional Publication*, **16**, 1–170.

Chamberlain, A. C. (1975) The movement of particles in plant communities. In *Vegetation and the Atmosphere. 1. Principles*, Monteith, J. L., ed., Academic Press, New York and London, 155–203.

Christophersen, N. and Wright, R. F. (1981) Sulfate flux and a model for sulfate concentrations at Birkenes, a small forested catchment in southernmost Norway. *Water Resources Research*, **17**, 377–89.

Christophersen, N., Seip, H. M. and Wright, R. F. (1982) A model for streamwater chemistry at Birkenes, Norway. *Water Resources Research*, **18**, 977–97.

Claridge, G. G. C. (1975) Automated system for collecting water samples in proportion to stream flow rate. *New Zealand Journal of Science*, **18**, 289–96.

Cryer, R. (1976) The significance and variation of atmospheric nutrient inputs in a small catchment system. *Journal of Hydrology*, **29**, 121–37.

Edwards, A. C. and Cresser, M. S. (1985) Design and laboratory evaluation of a simple fractionating precipitation collector. *Water, Air, and Soil Pollution*, **26**, 275–80.

Edwards, A. C., Creasey, J. and Cresser, M. S. (1984) The conditions and frequency of sampling for elucidation of transport mechanisms and element budgets in upland drainage basins. *Hydrochemical Balances of Freshwater Systems*, IAHS Publication No. 150, Oxford, 187–202.

Edwards, A. C., Creasey, J., Skiba, U., Peirson-Smith, T. and Cresser, M. S. (1985) Long-term rates of acidification of UK upland acidic soils. *Soil Use and Management*, **1**, 61–5.

Florence, T. M. (1982) The speciation of trace elements in waters. *Talanta*, **29**, 345–64.

Fowler, D. and Unsworth, M. H. (1974) Dry deposition of sulphur dioxide on wheat. *Nature* (London), **249**, 389–90.

Frere, M. H. (1971) Requisite sampling frequency for measuring nutrient and pesticide movement with run-off waters. *Journal of Agricultural and Food Chemistry*, **19**, 837–9.

Galloway, J. N. and Likens, G. E. (1978) The collection of precipitation for chemical analysis. *Tellus*, **30**, 71–82.

Georgii, H. W. and Lenhard, U. (1978) Contribution to the atmospheric ammonia budget. *Pure and Applied Geophysics*, **116**, 385–92.

Haines, B. L., Waide, J. B. and Todd, R. L. (1982) Soil solution nutrient concentrations sampled with tension and zero-tension lysimeters: Report of discrepancies. *Soil Science Society of America Journal*, **46**, 658–60.

Hansen, E. A. and Harris, A. R. (1975) Validity of soil-water samples collected with porous ceramic cups. *Soil Science Society of America Journal*, **39**, 528–36.

Harrison, R. M. and Pio, C. A. (1983a) Major ion composition and chemical associations of inorganic atmospheric aerosols. *Environmental Science and Technology*, **17**, 169–74.

Harrison, R. M. and Pio, C. A. (1983b) Size differentiated composition of inorganic atmospheric aerosols of both marine and polluted continental origin. *Atmospheric Environment*, **17**, 1733–8.

Harrison, R. M. and Pio, C. A. (1983c) A comparative study of the ionic composition of rainwater and atmospheric aerosols: Implications for the mechanism of acidification of rainwater. *Atmospheric Environment*, **17**, 2539–43.

Hewlett, J. D. and Hibbert, A. R. (1963) Moisture and energy conditions within a sloping soil mass during drainage. *Journal of Geophysical Research*, **64**, 1081–7.

Kadowaki, S. (1976) Size distribution of atmospheric total aerosols, sulfate, ammonium and nitrate particulates in the Nagoya area. *Atmospheric Environment*, **10**, 39–43.

Klein, M. (1981) Dissolved material transport – the flushing effect in surface and subsurface flow. *Earth Surface Processes and Landforms*, **6**, 173–8.

Lewis, W. M. Jr and Grant, M. C. (1978) Sampling and chemical interpretation of precipitation for mass balance studies. *Water Resources Research*, **14**, 1098–1104.

Mart, L. (1979) Prevention of contamination and other accuracy risks in voltammetric trace metal analysis of natural waters. *Fresenius Zeitschrift für Analytische Chemie*, **296**, 350–7.

Martin, R. P. and White, R. E. (1982) Automatic sampling of stream water during storm events in small remote catchments. *Earth Surface Processes and Landforms*, **7**, 53–61.

Mayer, R. (1971) Bioelement-Transport im Niederschlagswasser und in der Bodenlösung eines Wald-Ökosystems. *Gottinger Bodenkundliche Berichte*, **19**, 1–119.

Mayer, R. and Ulrich, B. (1974) Conclusions on the filtering actions of forests from ecosystem analysis. *Oecologia Plantarum*, **9**, 157–68.

Miller, J. D. and Miller, H. G. (1976) Apparatus for collecting rainwater and litter fall beneath forest vegetation. *Laboratory Practice*, **29**, 850–1.

Miller, H. G. and Miller, J. D. (1980) Collection and retention of atmospheric pollutants by vegetation. In *Proceedings of the International Conference on the Ecological Impact of Acid Precipitation*, SNSF Project, Norway, 33–40.

Montgomery, H. A. C. and Hart, I. C. (1974) The design of sampling programs for rivers and effluents. *Water Pollution Control*, **73**, 77–98.

Nagpal, N. K. (1982) Comparison among and evaluation of ceramic porous-cup soil water samplers for nutrient transport studies. *Canadian Journal of Soil Science*, **62**, 685–94.

Newson, M. D. (1981) Mountain streams. In *British Rivers*, Lewin, J., ed., George Allen and Unwin, London, 59–89.

Nicholson, I. A., Paterson, I. S. and Last, F. T., eds (1980) *Methods for Studying Acid Precipitation in Forest Ecosystems*, Institute of Terrestrial Ecology, Cambridge.

Nielsen, D. R. and Biggar, J. W. (1962) Miscible displacement III. Theoretical considerations. *Soil Science Society of America Journal*, **26**, 216–21.

Nihlgård, B. (1970) Precipitation, its chemical composition and effect on soil water in a beech and a spruce forest in south Sweden. *Oikos*, **21**, 208–17.

Nordin, C. F. and Dempster, G. R. (1963) Vertical distribution of velocity and suspended sediment, Middle Rio Grande, New Mexico. *United States Geological Survey Professional Paper* 462B.

Nurnberg, H. W., Valenta, P., Mart, L., Raspor, B. and Sipos, L. (1976) Applications of polarography and voltammetry to marine and aquatic chemistry. II. The polarographic approach to the determination and speciation of toxic trace metals in the marine environment. *Fresenius Zeitschrift für Analytische Chemie*, **282**, 357–67.

Patterson, C. C. and Settle, D. M. (1976) Sampling, sample handling, analysis. In *Accuracy in Trace Analysis*, NBS Special Publication No. 422, Lafleur, P. D., ed, US Department of Commerce, Washington DC.

Rainwater, F. H. and Avrett, J. R. (1962) Error inference in systematic-sample statistics in stream quality studies. *Journal of the American Water Works Association*, **54**, 757–68.

Reid, J. M., MacLeod, D. A. and Cresser, M. S. (1981) The assessment of chemical weathering rates within an upland catchment in north-east Scotland. *Earth Surface Processes and Landforms*, **6**, 447–57.

Reynolds, S. (1984) An assessment of the spatial variation in the chemical composition of bulk precipitation within an upland catchment. *Water Resources Research*, **20**, 733–5.

Richter, D. D. C., Ralston, C. W. and Harmes, W. R. (1983) Chemical composition and spatial variation of bulk precipitation at a coastal plain watershed in South Carolina. *Water Resources Research*, **19**, 134–40.

Sanders, T. G. and Adrian, D. D. (1978) Sampling frequency for river quality monitoring. *Water Resources Research*, **14**, 569–76.

Schneider, H. I. and Angino, E. E. (1980) Trace-element mineral and size analysis of suspended flood materials from selected eastern Kansas rivers. *Journal of Sedimentary Petrology*, **50**, 1271–8.

Schnoor, J. L., ed. (1984) *Modelling of Total Acid Precipitation Impacts*. Butterworth, London.

Skiba, U., Edwards, A., Peirson-Smith, T. and Cresser, M. (in press) Rain simulation in acid-rain research – techniques, advantages and pitfalls. In *Proceedings of a Symposium on Chemical Analysis in Environmental Research, Merlewood, 19–20 November 1985*, Institute of Terrestrial Ecology.

Strangeways, I. C. (1982) Instruments for mountainous areas. *Nordic Hydrology*, **12**, 289–96.

Unsworth, M. (1980) Dry deposition of gases and particles onto vegetation: a review. In *Methods for Studying Acid Precipitation in Forest Ecosystems*, Nicholson, I. A., Paterson, I. S. and Last, F. T., eds, Institute of Terrestrial Ecology, Cambridge, 9–15.

Walling, D. E. and Teed, A. (1971) A simple pumping sampler for research into suspended sediment transport in small catchments. *Journal of Hydrology*, **13**, 315–37.

Ward, R. C. (1975) *Principles of Hydrology* (2nd edn) McGraw-Hill, London.

Weyman, D. R. (1970) Throughflow on hillslopes and its relation to the stream hydrograph. *Bulletin of the International Association of Scientific Hydrology*, **15** (3), 25–33.

Weyman, D. R. (1974) Runoff process contributing area and streamflow in a small upland catchment. In *Fluvial Processes in Instrumented Watersheds*, Gregory, K. J. and Walling, D. E., eds, Institute of British Geographers, London, Special Publication No. 6, 33–43.

Whitehead, H. C. and Feth, J. H. (1964) Chemical composition of rain, dry fall out and bulk precipitation, California 1957–1959. *Journal of Geophysical Research*, **69**, 3319–33.

Wolff, R. G. (1967) Weathering of Woodstock granite near Baltimore, Maryland. *American Journal of Science*, **265**, 106–17.

# 5

○ ○ ○ ○ ○ ○ ○ ○ ○ ○ ○ ○ ○ ○ ○ ○ ○ ○ ○ ○

# Possible priorities in freshwater acidification research

At first glance the title selected for this final chapter may surprise some readers. Cautious authors tend to avoid predicting what will be, or should be, done in the future, if only because such material may all too quickly become out of date as achievement surpasses prediction. However, we felt that it was very appropriate at this point to look at the relevant research in the past and at present, with a view to briefly summarising progress and identifying gaps in our knowledge. An attempt may then be made to provide some pointers as to where progress might most usefully be made. We have attempted to be very concise, since important points often fade with excessive speculation.

## Water acidification and aquatic biota

The biological aspects of this topic are almost beyond the scope of this volume. Much information has been accumulated over the past 15 years on changes in biological populations associated with changes in water quality from surveys of diverse streams or lakes. A considerable body of knowledge is accumulating of the importance of speciation of elements, such as aluminium, which are associated with freshwater acidification, and upon inter-element effects on biota. Some evidence is accumulating upon the recovery of depleted populations after liming of acidified waters.

Relatively little information has been obtained, however, about the effects of pulses of high acidity in streams, even though these are known to occur quite naturally. What little evidence there is suggests that rapid flushing of invertebrate populations may occur. This may subsequently affect higher species. A strong case could be made for more biological episodic research, and the results should be coupled to modelled acidification pulse data to predict effects of severe (ten-year or worse) storms on biota. Work on recoveries from acidification pulses in streams should not be neglected.

## Vegetational effects

Much effort has been expended in medium-term investigations of the effects of vegetation upon precipitation chemistry. However, a disproportionate amount of this has been directed towards understanding the changes in throughfall chemistry compared to rainfall. Knowledge of the processes consequently occurring in the soil is still severely restricted at a quantitative level. More often than not discussion is confined to broad generalisations based upon knowledge of soil physicochemical processes built up over many decades. Where acid deposition is neutralised by vegetation, little is known from direct experiment about the subsequent transfer of $H^+$ to the rhizosphere and its effect upon rhizosphere soil. This issue is crucial to both soil and water acidification in the long term, and should be examined in depth, ideally in association with work upon root growth, soil structure and water drainage through and from soil.

The uptake of nitrate by moorland vegetation and sometimes by trees is well documented. The ultimate fate of such retained nitrate in upland catchments is only understood in broad process-level terms. Detailed quantitative work on nitrogen is sparse compared with that on sulphur, and the significance of nitrogen uptake as a consequence of interception by vegetation to long-term soil and water acidification should receive more attention.

A further aspect of vegetation effects which warrants serious consideration is the significant changes that may occur in neutral salt fluxes in relation to vegetational changes, whether due to natural or anthropogenic effects. Increased neutral salt inputs may reduce soil solution pH, shift ion exchange equilibria, and change the local geochemical weathering environment. This aspect should receive attention too, and not just the additional acid flux deposited. Total sulphate flux should be considered with respect to soil sulphate saturation and the associated increase in base cation leaching.

The rate of degradation of plant remains is a crucial factor in the soil/plant/water ecosystem. Most studies of acid deposition effects upon microbial activity have hinged around changes (or lack of changes) in respiration rates or enzyme activities, or around partitioning of energy between respiration and biomass accumulation, as assessed by the fate of a $^{14}C$ glucose amendment for example. Relatively little work has been done as yet upon the possible production of organic biological inhibitors or stimulators as a consequence of changes in precipitation chemistry, or indeed upon the effects of deposition upon the biodegradability of litter. Such topics are under investigation in our laboratories. Their importance stems from the significance of soil atmosphere carbon dioxide concentration and the possible importance of changes in organic matter accumulation to surface horizon pH.

## Hydrological pathway research

While some excellent progress has been made in our understanding of hydrology of upland catchments over the past two to three decades, most of this work, including the modelling aspects, has been inspired by a desire to explain climate/discharge relationships for a given catchment. Much less work has been conducted with a primary objective of being able to explain discharge acidity in terms of hydrological pathways in relation to fundamental soil processes. Simple conceptual hydrological sub-models, such as that discussed briefly in the previous chapter, are not transferable from one catchment to another. There is therefore a definite need for a new generation of more sophisticated models based upon precisely defined soil parameters. This will undoubtedly involve far more national and international collaboration than has hitherto been achieved. The end product, very widely applicable models for short and long-term predictive purposes, should be worth the effort required however.

A specific topic that needs clarification in this context is the influence of tree growth upon hydrological pathways. Knowledge is needed for a variety of forest types, with due consideration being given to the interacting climate, pollutant, topography and soil type effects. Such studies should be conducted not just with a view to improving our understanding at the fundamental process level, but also with a mind to developing sound cultural practices for minimising acidification of drainage waters under diverse conditions. Snow and frozen soil effects also deserve more research effort.

## Carbon dioxide effects

The natural buffering effect of carbon dioxide upon upland stream-water chemistry has been discussed in Chapters 2 and 3. While many studies have been made of rates of carbon dioxide production in agricultural soils, knowledge of generation rates under diverse climatic conditions in upland soils is much more sparse. This information would be essential to accurate water acidity modelling at the fundamental process level.

## Soil acidification and recovery rates

Soil pH is crucial to natural water pH in many circumstances. There is therefore a very real need to develop reliable long-term models for prediction of changes in soil pH with time throughout the soil profile. Such sub-models are again an important prerequisite to universal models for long-term predictive purposes. In the very few instances where soil pH has been shown conclusively to have been lowered as a consequences of deposition acidity, modelling of potential for recovery as a result of future reduced emission levels should be given equal priority.

## Some concluding remarks

It would be satisfying to be able to end with a few concise, definitive statements that summarised the whole acid deposition/water acidification scenario in a few well-chosen words. The best we can offer is to say that each catchment must be considered individually. There are undoubtedly some types of terrain where acid deposition can and does lead to drainage water acidification. There are some soil/plant/water ecosystems where acidifying pollutants in the atmosphere may significantly lower soil pH and that of the associated streams or lakes. But it must not be overlooked that the generation of very acidic, organic surface soils is a perfectly natural process in a leaching climate, typically occurring over *ca* 10 000 years or less from granite. Such soils will give very acidic waters, initially during episodes, but progressively for more and more of the time as acid peat accumulation increases. Acidifying pollutants may accelerate the process somewhat, but it would eventually occur anyway.

# INDEX